John Wilkins Williams

A Clean Skin : How to Get it and how to Keep it

Skin Diseases of Constitutional Origin : Their Etiology, Pathology, and Treatment

John Wilkins Williams

A Clean Skin : How to Get it and how to Keep it
Skin Diseases of Constitutional Origin : Their Etiology, Pathology, and Treatment

ISBN/EAN: 9783744763233

Printed in Europe, USA, Canada, Australia, Japan

Cover: Foto ©berggeist007 / pixelio.de

More available books at **www.hansebooks.com**

A CLEAN SKIN:

HOW TO GET IT AND HOW TO KEEP IT.

SKIN DISEASES OF CONSTITUTIONAL ORIGIN:

THEIR

ETIOLOGY, PATHOLOGY, AND TREATMENT.

BY

JOHN WILKINS WILLIAMS,
M.R.C.S.Eng.
OF ST. JOHN'S COLLEGE, OXFORD;
LATE HOUSE-SURGEON TO THE LOCK HOSPITAL, LONDON.

LONDON:
SIMPKIN, MARSHALL, AND CO.
1864.

PREFACE.

My endeavour in this little book has been to produce a manual on Constitutional Skin Diseases, which, while embodying the matured results of my own experience, shall at the same time reflect the most recent Continental researches in this important class of maladies. A good book in this department of medical science has long been a *desideratum*. That there are three or four most able and elaborate treatises on skin diseases in the English language, I am the first to admit; but I feel sure that I merely re-echo a general complaint in asserting that their bulk detracts sadly from their usefulness, and that they are scarcely on a level with the most recent researches made in this department, especially by the French Dermatologists. These defects I have endeavoured to avoid in the present volume, which I hope will be found convenient in size, sound in practice, and well 'posted-up,' as our American cousins say, in each branch of the subject of which it treats.

Neither in form nor in matter does it assert any claim to originality. It is based almost wholly on the systems of the modern French school of Dermatologists. And here I must take the opportunity of acknowledging, once for all, my deep obligations to these gentlemen, more especially to MM. Hardy, Bazin, Rollet, Diday, and Langlebert, whose valuable writings I have freely laid under contribution in compiling the present volume.

It only remains for me to state the grounds upon which I have thought fit to adopt the French in preference to the ordinary English system of classification,—that, namely, of Willan and Bateman.

English writers have for the most part classified these diseases according to the nature of the cutaneous lesion. They make an erythematous class, a papular, vesicular, pustular, &c., according as the skin affection manifests itself in the form of an erythema, a papule, vesicle, or pustule. In other words, they adopt as the basis of classification the outward and obvious character of an eruption. But these characters, it must not be forgotten, are merely accidental, and a classification so founded throws no light on the real or essential nature of a skin disease. Dealing merely with local and superficial manifestations, it loses sight of (or rather fails altogether to see) that recondite and subtle constitutional diathesis which lies at the bottom of the mischief, and which is the real enemy we have to grapple with.

Now the French system is the very reverse of all this. Attempting a higher generalization, it classifies skin diseases, not according to their obvious characters, but according to the constitutional diathesis in which they are supposed to originate. I say '*supposed* to originate' purposely, because the existence of two of these constitutional diatheses, viz. the Dartrous and the Arthritic, is a questionable hypothesis rather than an undisputed fact. Still this does not detract from the value of their system of classification. Hypothesis, though it does not constitute knowledge, is an invaluable instrument in helping us to obtain it. If hypothesis is to be discarded in medical science, especially in so obscure a branch of it as Dermatology, we shall never advance a single step towards placing it on a more rational, less empirical basis. This is the reason why I lean towards the French school. I do not receive as established truth the speculations which it has advanced in Dermatology. I regard them, not as arrows shot home to the mark, but simply as arrows aimed in the right direction. How near they have reached their aim, further research will one day show. In the meantime it is quite certain that the French are probing deeper into the etiology of skin diseases than are our own Dermatologists; just as the man, who draws his bow even at a venture, stands a better chance of hitting the mark than he who lets his arrows lie idle in the quiver.

10 Wimpole Street,
    Cavendish Square.
        Nov. 1861.

CONTENTS.

PAGE

Preliminary remarks on Health and Disease—The Constitutional disorders, in which skin-diseases originate, hitherto very imperfectly known but now being studied by the French Dermatologists with great prospect of success—General characters of these disorders—Their four great classes—Relative gravity—and distinguishing features of each class 1–5

DARTROUS ERUPTIONS

Meaning of the word Dartre—General characters of the Dartrous diathesis—and of Dartrous eruptions (1) of external skin, (2) of internal skin or Mucous Membrane—Three chief forms of Dartrous eruption, viz. Eczema, Pityriasis, and Psoriasis 6–8

I. ECZEMA—

Its general characters and successive stages—First stage—Second stage—Third stage—Résumé of the characteristics of each stage—Troublesome concomitant phenomena—Eczema the type of a Dartrous eruption 9–13

Varieties of Eczema.

(a) According to its Aspect—

Eczema Simplex — Rubrum — Rimosum — Impetigo — Pityriasis—Lichen and its allied form Prurigo—their general characters and chief varieties, viz. Lichen Simplex, Circumscriptus, Agrius, Inveteratus, &c. 13–19

(b) According to its Form—

Eczema Diffusum and Circumscriptum—Varieties of the latter, viz. Eczema Figuratum, Larvale, Nummulare . . 19–20

(e) According to its Situation—
Eczema Pilare—of Face and Ears—Breasts—Navel—Genitals (1) in the female, (2) in the male—Perinæum and adjacent parts—Lower limbs—Hands and Feet 20–25

Acute Eczema. Treatment of the First stage, locally and internally—of the Second stage, locally and internally—of the Third stage (and Chronic Eczema)—α. Internal: Tonics, Cod Liver Oil, Arsenic, Sulphur, Cantharides, &c.—β. Local: Ointments, Counter-irritation, Baths.—γ. Hygienic 25–31

II. PITYRIASIS—
Its general characters—and different varieties, viz. Pityriasis Rubra, Circinata, and Pilaris—Treatment of each variety (1) medicinal, (2) hygienic 32–33

III. PSORIASIS—
Its favourite seats—general characters—and different varieties, viz. Psoriasis of the Head, Nails, Hands and Feet, Face, Prepuce—Temperament and Sex most liable to Psoriasis—Treatment (1) Internal: Arsenic, Cantharides, Copaiba. (2) External: Baths, Ointments. (3) Hygienic 34–39

ARTHRITIC ERUPTIONS

Arthritis—its general characters and several stages.
Arthritic Eruptions—their general characters as to Pain, Seat, Form, Colour, Secretion, Complexity, Progress, and Behaviour with Remedies . . 40–42

Varieties of Arthritic Eruptions recognized by M. Bazin.
 I. The PSEUDO-EXANTHEMATOUS—
  Erythema Nodosum, Urticaria Febrilis, Cnidosis, Herpes and Zona. Pemphigus 43–45
 II. The DRY—
  Intertrigo, Couperose, Pityriasis, Psoriasis, Lichen, Acne . 45–46
 III. The MOIST—
  Eczema, Mentagra, Ecthyma and Furunculus . 46–48

Preventive treatment of Arthritic eruptions—Curative treatment (1) internal: Alkalies, Alkaline Spas, Colchicum, &c. (2) local: Emollient Powders, Oil of Cade, Benzine, M. Bazin's Formulæ, Morphia . . . 48–51

SYPHILITIC ERUPTIONS

Three distinct kinds of *primary* venereal disease, viz. :

A. Gonorrhœa—

    Its History — Symptoms — Cause—Stages — Treatment—Sequelæ troublesome, especially Gleet—Causes and varieties of Gleet, and its treatment in the male and in the female 52–58

B. Non-infecting or Soft Chancre

    Its characters—Treatment by Caustics and by milder means Points of difference between it and 58–61

C. Infecting, Hard, or Hunterian Chancre.

    Characters and treatment of the latter . . 62–64

Secondary Venereal affections.

  (a) Syphilitic Vegetations—

    Condylomata; their physical characters, favourite localities and modes of origin

    Excrescences; their nature and treatment . 64–65

  (b) Syphilitic skin-diseases—

    Their Colour—Form—Painlessness—Multiformity—Products—Seat and concomitant phenomena—M. Hardy's Classification . . 66–68

    I. PIGMENTARY.

    II. EXANTHEMATOUS: Syphilitic Roseola — its characters, progress, and concomitant phenomena.

    III. VESICULAR—its varieties, viz. Eczematous, Varioliform, Herpetiform.

    IV. PUSTULAR—its varieties, viz. Acneiform, Ecthymatous, Pustulo-Crus-

    V. PAPULAR. [taceous.

    VI. BULLOUS—Pemphigus Neonatorum

    VII. SQUAMOUS—its varieties, viz. Lepra, Psoriasis, and Callosity.

    VIII. TUBERCULAR—its varieties, viz. Aggregated, Disseminated, Perforating, Serpiginous 68–79

Treatment of Syphilitic skin-diseases—

    Indiscriminate administration of Mercury a practice well-nigh obsolete, and open to grave objections—Non-Mercurial treatment the safest, but in certain cases insufficient—Therapeutic indications to be derived from the nature of the First, and of the Second eruption—How long Mercury should be given—Modes of obtaining its influence upon the system—(1) by the mouth; (2) by inunction; (3) by fumigation—Superiority of the last method—Treatment of Tertiary symptoms—Diday's views of Tertiary Syphilis—General Hygiene: Food, Air, Sleep, Exercise, Continence, Sea-bathing &c. . . . 79–88

Contagiousness of Secondary Syphilitic Lesions—

    Doctrine of the Contagion of Secondary Syphilis now well established—Evidence by which it is supported—Difference between results of infection from secondary and from primary lesions, in respect of their incubation, physical characters, and virulence . . 88–90

SCROFULOUS ERUPTIONS

PAGE

Scrofula—Its general characters and influence upon the physical conformation; physiognomy; stature; physical, mental, and moral temperament; and sexual impulse—Phenomena of its different stages—Scrofulous skin-diseases—their distinguishing characteristics . . 91–95

Principal varieties of Scrofulous Eruptions.

I. Seated in the superficial layers of the skin—
Chilblains—their cause—Treatment, locally and constitutionally—*Acne* and its varieties—preventive treatment—curative treatment—local applications 95–99
Molluscum 99–100

II. Seated in the deeper layers of the skin—
Lupus and its varieties. viz. Erythematous, Non-Exedens, Exedens—Treatment of Lupus internally and locally 100–102
General characters of the deep cutaneous lesions in Scrofula . 103–104

III. Seated in the Mucous Membranes—
Scrofulous affections in these membranes either Catarrhal or Eruptive. The *Catarrhal*—their peculiar characters—and effects on the mucous membrane of the Eye, Nose, Ear, Fauces, Genito-Urinary Canal, Lungs and Intestines. *The Eruptive*—their treatment . 104–109

General Treatment of the Scrofulous diathesis:

A. Hygienic—
Light and Air—kind of climate desirable—Ablution and friction of Skin—Sea-bathing—Open-air exercise—Clothing . . . 109–111

B. Dietetic—
Food—Alcohol 111–112

C. Medicinal—
Mal-Assimilation of food the great evil to be remedied—No known specific against Scrofula—Therapeutic value of the Vegetable Bitters, Alkalies, Acids, &c.; of Purgatives; of Cod Liver Oil, Cream, Iodine, Steel, Bark, &c. 112–114

SKIN DISEASES

OF

CONSTITUTIONAL ORIGIN.

How rare a blessing is uninterrupted and perfect health! How few there are who have not to endure, at some time or other of their lives, more or less of physical suffering! Health depends on so even a balance, so exact an adjustment, of the vital forces—while the influences, both from within and from without, tending to disturb this balance are so numerous and so unceasing in their operation, that the wonder is not how or why we fall ill, but how in the world we manage to keep well. A moment's reflection will justify such an observation. To say nothing of the countless disturbing influences outside ourselves (such as the evil results of mental and physical exhaustion, bad food, bad air, heat and cold, contagion, mechanical violence, &c.) we have only too many evil influences at work within. Not a few of us, being born of sickly parents, come into the world already diseased; or, if not actually diseased, with more or less predisposition to disease. Such predisposition may haply lie dormant, but more often is only too ready to burst forth under circumstances favourable to its developement. Again, even if we have the good luck to be constitutionally sound as regards

Preliminary remarks on health and disease.

any hereditary taint, we cannot avoid the strong tendency there is to particular disorders at particular periods of life.

The liability to disease being so great, medical philosophers from time to time have set themselves to enquire what disease really is. The speculative nature of the human mind is for ever tempting us on to the attainment of higher truths. In every progressive science the grand motto of its devotee is Excelsior! Not content with the contemplation of individual facts, he yearns to know what those facts mean when put together. So it is with Medicine. The nature of individual *diseases* has been and is being studied with so much success, that medical men have long been trying to grasp the essence of *disease*. But all such attempts have as yet proved futile. Hippocrates, Galen, Sydenham, and a host of medical authorities, have tried hard to define the word, but without elucidating its real nature. Their definitions are true as far as they go, but they do not go far enough. Their general tenor has been to describe disease as some morbid alteration of structure or function or both. Definitions of this kind, dished up in sonorous sesquipedalian words, may satisfy unthinking minds, but will never satisfy those who think seriously what words mean. They lighten one's darkness just about as much as the well-known reply given to the sick Grandee, who on asking his doctor 'Why opium sent people to sleep?' was told it was 'Because it has a soporific tendency.' The truth is that, in the present state of our knowledge, a real definition of disease is impossible. Not having sufficient data to go upon, we can as yet merely say what disease does, not what it is. Just as we know the wind to be astir by its effects—that is to say by feeling its breath and hearing its sound—so we recognise and study disease by its phenomena or symptoms. It is likewise against the symptoms of a disease that our treatment is generally directed, not against the disease itself. The latter is often too obscure or too subtle for us to fight with, and its symptoms are the only tangible points we can lay hold of.

To no class of diseases do these remarks apply with greater force than to such diseases of the skin as are of constitutional origin. In skin diseases of local origin (Scabies for instance and the different kinds of Tinea), the cause is superficial and within easy reach. By killing the parasite we put an end to the disease. But in those of constitutional origin, the cause is generally so deep-seated and so subtle as to elude all our present means of detection, and the treatment is of course proportionately empirical. The *fons et origo mali* is here some change in the constitution of the blood, rendering it unfit for the healthy nutrition of the cutaneous system. At present these different blood diatheses, which manifest themselves in different forms of skin disease, are most imperfectly known. They have hitherto eluded the keen search even of those expert detectives, the microscope and the test-tube. It is not surprising that a class of diseases, whose treatment is attended with so much empiricism and uncertainty, should have been comparatively shunned as subjects of investigation by medical men. Their etiology and pathology have, in fact, been advanced scarce a single step since the close of the last century, when Willan and Bateman founded their well-known system of classification. This system reigned almost supreme both in England and on the Continent, till it found a rival in that of the modern French Dermatologists, of whom we may take MM. Bazin and Hardy as the representatives. For the reasons stated in my preface and which I need not here repeat, I am

The constitutional disorders, in which skin diseases originate, hitherto very imperfectly known,

but now being investigated by the French Dermatologists with great prospect of success. convinced of the superiority of this latter system of classification, and that it affords the only means by which our knowledge of this department of medical science is likely to be advanced. It is, therefore, the system of which I shall endeavour to give an outline in the following pages.

General characters of these disorders. Constitutional skin diseases may be acute or chronic, inflammatory or non-inflammatory, continuous or intermittent, contagious or non-contagious, characterized by an assemblage of various morbid phenomena, from which none of the organic structures can be pronounced exempt. Further, diseases of this class may be either hereditary, acquired, or dependent on constitutional predisposition, and are capable of subdivision *Their four great classes;* into four great classes—viz. Herpetic or Dartrous, Syphilitic, Scrofulous, and Arthritic. The skin is usually the part in which they first make their appearance; they next attack the lymphatic system, which is the favourite seat of Scrofula and Syphilis, while in Dartrous and Arthritic affections the lymphatic is attacked subsequently to the tegumentary system. At a later period the nervous system, cellular tissue, serous membranes, bones and viscera, may fall victims to the ravages of these disorders. The constitutional diseases in which cutaneous eruptions take their origin are not all, however, equally grave. M. Bazin very properly places them *Relative gravity;* in the following order, commencing with the least and terminating with the most serious:

1. HERPES or DARTRE.
2. ARTHRITIS.
3. SYPHILIS.
4. SCROFULA.

Before proceeding, it will be desirable to give a brief *résumé* of the principal characteristics of each class.

Cutaneous and visceral affections are incidental to all four, but the lymphatic system is apt to be affected in two only— viz. Scrofula and Syphilis. *and distinguishing features.*

The subjects of Dartre are liable chiefly to neuralgic, those of Arthritis to rheumatic pains, while in Syphilis the pain is usually of a mixed character.

Scrofula, Syphilis, and Arthritis, attack chiefly the bones and joints; Dartre, on the other hand, has a predilection for the skin.

Dartrous and Arthritic eruptions rarely penetrate beneath the superficial layers of the skin, are hardly ever ulcerative, are generally attended with heat and irritation, and leave behind them only temporary discoloration. Scrofulous and Syphilitic eruptions, on the other hand, are wont to implicate the deep cutaneous structures, are often ulcerative, are seldom attended with heat or irritation, but usually leave behind them strongly marked scars.

While hereditary Scrofula and Syphilis in the majority of cases commence in infancy, Dartrous and Arthritic affections do not usually manifest themselves till middle life or old age.

In Scrofula the venereal appetite is either very strong or very weak, more often the former; and, owing to the irritability of the genito-urinary mucous surfaces, seminal emissions are of frequent occurrence. In Dartre and Arthritis the animal passions are usually strong; but when, as often happens, the eruption attacks the generative organs, involuntary emissions are apt to ensue. In either case these latter, if allowed to occur with undue frequency, are sure to entail diminution of sexual power.

Lastly, all four diseases either terminate favourably after a longer or shorter period of treatment, or are prolonged until the patient's death.

DARTROUS ERUPTIONS.

Meaning of the word 'Dartre.' BEFORE I proceed to discuss the different varieties of this class of eruptions, some explanation is needed of the term *Dartre* which will doubtless sound new in the ears of English readers. *Dartre* is an old French word used by the French Dermatologists to designate a constitutional taint, which manifests itself in various forms of cutaneous affections; the latter being prone to recur from time to time, and the patient in the interval enjoying perfect health. In this sense it comprises the different eruptions described by the Willanists as Eczema, Psoriasis, Impetigo, &c. To those accustomed only to the ordinary English classification of skin diseases, it may seem strange to regard squamous, papular, vesicular and pustular eruptions as members of the same family. But it is quite conceivable that the same morbid process which, when limited to the epidermis, gives rise to a scaly eruption, may, when it involves the dermis itself, appear as a papular, vesicular, or pustular eruption, according to its intensity. Such a classification at any rate rests on a far more comprehensive basis than does our common English one. Looking beyond the mere outward skin disease, it aims at discovering the constitutional affection in which the said disease originates.

General characters Persons affected with the Dartrous diathesis, although to outward appearance possessing all the attributes of perfect

OF CONSTITUTIONAL ORIGIN.

health, are yet in a peculiar state which is not sound condition. Their skin is habitually dry and, although perhaps free from eruption, is so sensitive as to become chafed and irritated on the slightest provocation. Strong coffee, highly-spiced dishes, alcoholic indulgence, will sometimes in a Dartrous subject cause general erythema; or again, the folds of a garment, the application of a plaister, or some such trifle, will set up a local eruption, slight perhaps in itself, but still sufficient to indicate a peculiar constitutional taint—in other words, the existence of concealed mischief, needing only longer time or greater provocation to declare itself under its true colours. *of the Dartrous diathesis,*

Dartrous eruptions may, and often do, comprise a mixture of the elementary cutaneous lesions—as for instance in Eczema, which often exhibits vesicles, pustules, scales, &c. simultaneously and on the same patch of skin. They are generally slight, superficial, and mobile—that is to say, rarely limiting themselves to a single spot. Their tendency is to recur and spread, sometimes by gradual invasion of the adjacent sound skin, sometimes by attacking simultaneously or successively portions of skin more or less distant from each other. Another peculiarity is the symmetry which they affect in the same situation on either side of the trunk. They are generally accompanied with heat and irritation, which are apt to be so intense as to cause sleepless nights and exhaust the nervous system. They give rise to excoriations wide in extent but shallow in depth, and leaving behind them at the worst only slight and transient discolorations. This latter, as before observed, is an important point of diagnosis between these affections and those of Scrofulous or Syphilitic origin. They are non-contagious. They attack all ages, both sexes, and persons of all temperaments. In the majority of cases they last for months and even for years, with longer or *and of Dartrous eruptions (1) of external skin,* shorter periods of intermission. It is this tendency to relapse which renders their treatment so unsatisfactory; indeed to guarantee permanent cure of a Dartrous eruption is more than any medical man can honestly promise. Each time it disappears, the local affection alone is subdued, not the constitutional diathesis, which merely slumbers until again evoked by some fresh disturbing cause. Hygiene, diet, and judicious medical treatment will aid the immediate cure of the eruption; nay more, they will probably defer its relapse, but they cannot altogether prevent its recurrence.

(2) of internal skin or mucous membrane.

The mucous membranes or internal skins are also liable to these eruptions, though in a less degree than the skin of the external surface. Thus Dartrous affections of the face will attack sometimes the mucous membrane of the eye, sometimes that of the auditory canal of the ear; and according to their intensity will interfere more or less with the functions of these organs. The urethra in the male and vagina in the female are also open to attack, and so may become the seat of troublesome discharges. The Dartrous diathesis seems to have a greater predilection for the mucous membrane of the respiratory than the digestive apparatus. Sometimes, however, it attacks the latter and causes visceral derangement in the shape of dyspepsia and diarrhœa, which may either alternate or coincide with the cutaneous affection.

Three chief forms of Dartrous eruption.

Skin diseases of the Dartrous character resolve themselves into three main groups—viz. Eczema, Pityriasis, Psoriasis. Other skin diseases exist (Lichen and Impetigo for instance) which, although of the Dartrous type, are not of sufficiently distinctive importance to entitle them to a separate division, inasmuch as they are merely modifications of the largest and most important division, Eczema.

ECZEMA.

Eczema, the most common of all cutaneous affections, is designated by Alibert 'Herpes Squamosus Madidans'—a happy expression, imparting a good idea of the scaly appearance of the spot affected and of the secretion with which its surface is ordinarily moistened. To allot Eczema its proper place in the group of skin diseases is easy enough, but to give a true description and intelligible definition of the various forms under which it appears is a matter of considerable difficulty.

Eczema.

Eczema perhaps may be appropriately described as an affection of the cutaneous and mucous surfaces, which at its commencement is characterized by the appearance of either exanthemata, vesicles, scales, or papules. The next feature is the presence of a serous or sero-purulent secretion, more or less abundant in quantity, which, as it dries, forms a successive series of crusts; each crust being gradually loosened and finally detached by the accumulating secretion beneath it. This definition will be found to comprise all the various forms which Eczema assumes—Eczema with sero-purulent secretion as well as dry Eczema; Impetiginous Eczema of the scalp as well as Fissured Eczema of the hands; Pityriasis, which is merely the usual mode of termination of Eczema, as well as Lichen, which is nothing more than a papular variety. It, in fact, enables us to make a comprehensive classification of all those affections which are nothing but the same disease either during successive periods or under different aspects, varied according to situation or to inherent constitutional predisposition. Still further, however, to facilitate the description of Eczema, I shall divide its progress into three distinct stages.

Its general characters,

and successive stages.

The first phenomenon is a redness more or less extensive upon

First stage.

which sometimes vesicles, sometimes vesico-pustules, sometimes simple cracks in the epidermis, are noticeable. The vesicles appear in the form of little eminences slightly raised above the surface of the skin, closely grouped together, and perfectly transparent as if filled with clear serum. They continue in this state rarely more than from thirty-six to forty-eight hours, so that in this early stage they do not often fall under the surgeon's notice. Sometimes they are packed together so closely as to coalesce and form a large bulla, similar to those formed in Pemphigus. In some rare cases of chronic Eczema the vesicles disappear, their serous contents being reabsorbed; but in the majority of cases the vesicles break, either spontaneously or by the contact of the nails in scratching. Their rupture leaves little superficial ulcerations or in common parlance 'raws,' moistened with a transparent sero-plastic secretion which stains the linen and makes it adhere to the part affected. This secretion dries on the surface and forms delicate crusts of a yellow or grey colour. Again, instead of vesicles pustules or vesico-pustules may present themselves; but these are, after all, only vesicles in which pus has taken the place of serum. These pustules appear in groups and have but a brief existence. Their liquid contents also form crusts, but more thickened, unequal, irregular, and of a deeper yellow or green colour than those before mentioned. In some rare cases Eczema begins with cracks or fissures in the epidermis, forming sinuous lines crossing each other in all directions; these, again, discharge a sero-plastic fluid which presents the same characters and goes through the same stages as in the other varieties of Eczema. This latter form is, however, far from common; and exanthematous patches, vesicles, and pustules must be regarded as the distinctive features of Eczematous eruptions.

OF CONSTITUTIONAL ORIGIN.

More frequently the affection has entered upon the second stage before it is presented to our notice. Vesicles and pustules have now disappeared and given place to other symptoms, more permanent and not less marked than those of the first stage. The excoriations which now appear are always superficial, sometimes isolated, sometimes by their union involving a large extent of surface. They secrete a plastic fluid which stiffens and stains the linen, and is serous or purulent according as the eruption began with vesicles or pustules. Further, as this fluid varies with the nature of the preceding lesion, so it determines the character of the succeeding crusts; these latter being thin and even if the fluid has been merely serous, but thick and uneven, if it has been purulent. These crusts, then, are the distinctive feature of the second stage of Eczema. At last they fall off either spontaneously or by the effect of treatment, and leave a red surface covered with excoriations or superficial ulcerations. These latter secrete a fresh instalment of plastic fluid, which forms crusts as before and shares the same fate as its predecessors. The duration of this stage is longer than the first; sooner or later, however, the inflammation subsides and the disease passes into the third stage.

Second stage.

All the crusts have now fallen off and the surface they covered is left red and dry; all secretion ceases; a sort of varnish seems to coat the part and helps to form small brilliant delicate lamellæ, which become white, opaque, and finally break away, to be succeeded by others finer in quality and less abundant in quantity. The scales become furfuraceous and after a longer or shorter time are replaced by a kind of farinaceous dust, which again diminishes until it ceases to reappear, and the patient is finally cured.

Third stage.

These are the three stages of Eczema. I must not, however,

omit to observe that the disease for a while generally leaves traces behind it, either in the varnished glowing appearance, or else in the thickening, roughening, and wrinkling of the skin, so well marked in the Lichenous variety of the complaint. Before proceeding further, it will be well briefly to recapitulate the main characters of each of the three stages of Eczema.

Résumé of the characteristics of each stage. Vesicles, vesico-pustules, fissures, and erythema of that portion of skin in which these alterations are developed, are the lesions of the first stage.

The notabilia of the second are secretion of serous or seropurulent fluid; crusts, varying in colour from golden-yellow to brown; superficial ulcerations and an erythematous condition of skin.

Shedding of the crusts; thickening, wrinkling, and burnishing of the cutaneous surface; these form the chief features of the third stage.

Troublesome concomitant phenomena. These general phenomena are accompanied by certain others equally deserving attention;—(1.) inflammation and swelling of the subcutaneous cellular tissue, which sometimes proceed to such an extent as to form little superficial abscesses; (2.) feeling of heat and irritation. The former, while it lasts, demonstrates the activity of the disease and, even when the surgeon deems the eruption cured, indicates by its persistence the approach of a speedy relapse. When, on the other hand, the feeling of heat subsides, a favourable issue of the disease may be reasonably expected. The latter symptom increases towards evening, causing sleepless and exhausting nights. Not unfrequently it is attended with a feeling of tension and burning, and so irritates the patient that he cannot refrain from scratching the part, thus aggravating the mischief and prolonging its duration.

OF CONSTITUTIONAL ORIGIN.

Eczema then may fairly be regarded as the type of a Dartrous eruption, possessing in a marked degree all the salient features of this class of skin diseases. It generalizes itself, either abruptly making its appearance in distant parts, or else spreading by continuity of tissue to adjacent parts, but scarcely ever involving at once the whole surface of the body. It is usually symmetrical; it has a tendency to invade the mucous membranes, especially those of the eye, mouth, tongue, nose, urethra, and vagina, causing in the two latter situations discharges of a most obstinate and intractable character. Its extension either to the anus or prepuce is an event of no unfrequent occurrence and constitutes a further source of most serious discomfort to the patient. All these eruptive affections of the mucous membranes may alternate or coincide with the primary affection of the cutaneous surface. *Eczema the type of a Dartrous eruption.*

The varieties of Eczema are numerous—far too numerous to receive a detailed account within the compass of this little volume. To do them full justice would necessitate a work of far more plethoric dimensions. I shall therefore content myself with sketching their outlines, omitting to fill in the details; for which purpose they may be conveniently divided into three groups according to their aspect, form, and situation. *Its varieties.*

Varieties of Eczema according to its Aspect.

This group may be sub-divided as follows:—

1. *Eczema Simplex.*
2. *Eczema Rubrum.*
3. *Eczema Rimosum.*
4. *Impetigo.*
5. *Pityriasis.*
6. *Lichen and Prurigo.*

Eczema Simplex is but a transient form of the complaint, *Eczema Simplex.*

generally ushered in by slight febrile symptoms, headache, loss of appetite, with heat and irritation in the parts about to become the seat of the disease. An erythematous blush next makes its appearance, and is soon followed by an eruption of vesicles or vesico-pustules filled with transparent serum. Before long these pass away, generally without breaking, and simultaneously with their resolution the local and constitutional symptoms disappear. The vesicles are replaced by a layer of small delicate shining scales, such as make their appearance towards the termination of an attack of Eczematous Pityriasis. If the vesicles are broken instead of being re-absorbed (as is frequently the case when the eruption involves a large surface), then small crusts form which soon fall off, and a cure speedily ensues. Eczema Simplex rarely lasts more than ten days; and as a general rule the larger the area of skin involved, the more rapid the cure of the disease.

Eczema Rubrum.

Eczema Rubrum bears a close resemblance to the preceding variety; in fact, it merely repeats its characters in an exaggerated degree. Its onset is nearly always attended with constitutional disturbance. Indeed these preliminary constitutional symptoms (malaise, headache, loss of appetite, &c.) accompany the onset of almost all eruptive diseases and in some rare cases—especially in very early life and when accompanied with cerebral disturbance—declare themselves with such severity as to endanger and even destroy life. The eruption in E. Rubrum consists in the simultaneous developement, either on several distinct parts or on one large surface, of red patches preceded by a sensation of heat and irritation. Next appear vesicles which have a great tendency by their coalescence to form bullæ. The fluid which they exude hardens into thin crusts; and these, when detached, leave superficial ulcerations which soon heal. Whatever the peculiarities of an individual case, the eruption pursues a tolerably regular course and finally terminates by desquamation. An important feature in this variety of Eczema is the swelling, at the part affected, of the subcutaneous cellular tissue, indicating that the vascular and highly-organized structures of the dermis are implicated. In this stage of its progress it sometimes bears no little resemblance to Erysipelas, but between the two diseases points of difference may be detected so many and so plain that ordinary caution would never confound the one with the other. E. Rubrum, though generally an acute disease of two or three weeks' duration, will sometimes localize itself in a particular spot—especially the face, hands, or genital organs—and there become tediously chronic.

Eczema Rimosum can always be recognised by the number of zigzag fissures which discharge a serous fluid similar to that contained in ordinary Eczematous vesicles. This secretion dries and forms crusts which fall off to be succeeded by others. The fissures gradually diminish in size and depth, the secretion becomes less plentiful, the crusts cease to form, and a cure slowly takes place, leaving the patient prone to frequent relapses. The importance of this variety depends not so much on the local affection as on its tedious progress and tendency to recur, which constitute it one of the most troublesome forms of Eczema. Its favourite localities are the flexures of the limbs and the different outlets of the body, especially the lips and the anus. In the latter situation it becomes a source of constant misery to the patient, giving rise to itching, burning, and painful defæcation.

Impetigo (the *Melitagra Flavescens* of Alibert), though essentially a pustular affection, has always been placed by the

<small>Eczema Rimosum.</small>

<small>Impetigo.</small>

16 SKIN DISEASES

French authorities in the Eczema tribe, as being the same malady under a slightly different, more aggravated form. It commences by little raised pustules beneath the epidermis, containing a thicker fluid than the common Eczematous vesicles. These pustules usually burst and by their desiccation form thick, wrinkled, honey-like crusts of a yellow or greenish-brown colour. Underneath is a raw surface, which is soon covered by fresh crusts composed of epidermis and dried sero-purulent fluid. These in their turn become detached and leave a red dry desquamating surface, which soon regains its normal appearance without any resulting disfigurement in the way of cicatrization. The disease is ushered in by symptoms of constitutional disturbance, together with heat and irritation in the part about to be affected.

Eczema and Impetigo then present the same lesions, phases, periods of eruption, secretion, desiccation and desquamation. Their only difference consists in the intensity of the inflammation, which is greater in Impetigo than in ordinary Eczema and therefore developes pustules instead of vesicles.

Pityriasis. *Pityriasis* exists under many forms; but the Pityriasis styled *communis* because of its frequency, *alba* because of its colour, *simplex* because of its favourable termination, is the variety which we have to notice at present. It may occur as an independent eruption, but more frequently it is merely the sequel or mode of termination of Eczema. It is characterized by a quantity of little white epithelial scales, resting on a surface somewhat redder than usual; it is attended with only slight irritation and yields readily to suitable treatment.

Lichen and Prurigo. *Lichen*, with its allied form *Prurigo*, usually commences by a feeling of heat and irritation so intense that patients cannot refrain from scratching; more especially as the act, besides affording temporary relief, is accompanied by a positive sensation of pleasure. There soon appears an eruption of little papules, closely grouped, full, but containing neither serum nor pus. These are frequently intermixed with vesicles or vesico-pustules, which get broken either by friction or scratching; and the result is a collection of papules and vesico-pustules, some intact, others broken and covered with yellow crusts. These crusts fall off and are renewed from time to time, until at last the skin remains clean, but presenting alterations which may be regarded as characteristic of these two forms of Eczema; viz. roughness, thickening, and exaggeration of its folds. The irritation is sometimes so great as to cause sleepless and exhausting nights with the natural sequelæ of nervous prostration, loss of flesh and strength. The patient cannot for his life resist scratching. Repose is out of the question. No sooner does he get warm in bed than he has to leave it and walk about his room, in the vain hope that exposure to air may allay the irritation. When this condition of things has continued a long time and still no relief ensues, it has been known to set up cerebral disturbance and drive the patient, under the mental excitement so induced, to seek relief by suicide from his unceasing and unbearable misery.

Lichen and Prurigo at first are usually limited to a single spot, but, like all dartrous affections, have a tendency to spread. All parts of the body are susceptible, but the back and sides of the neck, the anterior surfaces of the thighs, the hands, feet and genital organs, are their favourite localities. Of the numerous varieties of Lichen, the following are the most common and therefore the most useful to know.

Lichen Simplex, which is the least complicated form, yields

The general characters, and chief varieties, viz.

Lichen Simplex.

18 SKIN DISEASES

readily to remedies judiciously applied; but, owing to the strong tendency which Lichenous affections have to recur, treatment should be continued for a considerable time after the patient is apparently cured.

Lichen Circumscriptus. The peculiarity of this form is its appearance in circular patches about the size of a crown-piece, consisting of slightly elevated, closely grouped papules and rough scales, which finally disappear leaving a clean surface underneath. Another feature of this variety is the gradual subsidence of the eruption at the centre of the patch, while along the circumference it is spreading with unabated activity. Hence its liability at first glance to be mistaken for a form of Ringworm.

Lichen Agrius is a mixture of Eczema and Lichen, presenting by their simultaneous developement the features of both affections. The diagnosis is therefore sometimes difficult, but practically this is of no consequence since the same treatment is applicable to both. The symptoms are heat and irritation, coexistence of papules and vesicles, which latter break and form crusts—in fact, the usual assemblage of symptoms which characterize ordinary Eczema. This eruption in its immediate effects is amenable to treatment, but the tendency to relapse is so great that from four to six months are usually needed to effect a permanent cure.

Lichen Inveteratus is another form, important not so much on account of the intensity of the eruption as for the peculiar alteration in the integument, which becomes so thick and wrinkled that it is very difficult to take a piece between the finger and thumb, or to make it move over the subcutaneous tissue. It is essentially chronic and most obstinate in its recurrence as well as in its defiance of remedies.

The subjects of this disease present withered scaly thickened wrinkled skins, as if old age, care, and trouble had heavily impressed their stamp upon them.

Dermatologists have taken much pleasure in naming and describing several other varieties of Lichen, e. g. *Lichen Urticatus*, which is an Erythema; *Lichen Tropicus*, a mixture of Erythema and Urticaria peculiar to hot climates; *Lichen Lividus*, dependent on a cachectic condition of the patient; *Lichen Pilaris*, which is merely a Pityriasis affecting parts where the hair grows; *Lichen Podicis*, a most obstinate affection, rarely confining itself to the margin of the anus but generally extending from thence along the perinæum to the genitals, and giving rise to intense irritation and, too often, unchaste manipulation of these parts. Other varieties of Lichen.

Lichen and Prurigo, though attacking all ages and both sexes, still seem to have a predilection for subjects of a nervous temperament. The same causes which arouse Eczematous may produce Lichenous affections; change of seasons, excesses at table, over-indulgence of the emotions and over-exercise of the mental faculties, certain trades and professions—these are some of the main predisposing and exciting causes of this unpleasant disease.

Varieties of Eczema according to its Form.

Less important and less numerous than the preceding, these varieties of Eczema differ from the typical eruption only in form and extent of surface. They comprise *Eczema Circumscriptum* and *Eczema Diffusum*. The latter is, as its name indicates, diffuse, but presents no other feature calling for notice. The former exists under three modifications, Eczema Diffusum.

20 SKIN DISEASES

Eczema Circumscriptum and its varieties.

viz. *Eczema Figuratum, Eczema Larvale,* and *Eczema Nummulare.*

Eczema Figuratum presents the vesicles of Eczema and the pustules of Impetigo. It attacks the trunk, limbs, and face, and is characterized mainly by the irregularity of its borders; hence the propriety of its epithet.

Eczema Larvale attacks only the face, but the disfigurement which it there produces is sometimes so great as completely to disguise the features; hence the epithet, 'larvale' from 'larva,' a mask.

Eczema Nummulare consists of sharply-defined circular patches, resembling the marks which a coin would leave if impressed on the skin. These patches are generally eight to ten in number and are apt most obstinately to resist every species of treatment, local and constitutional.

Varieties of Eczema according to its Situation.

In this, the last division of the Eczema family, there are several varieties which merit careful study. They are:—

1. *Eczema Pilare.*
2. *Eczema of the Face and Ears.*
3. *Eczema of the Breasts.*
4. *Eczema of the Navel.*
5. *Eczema of the Genitals.*
6. *Eczema of the Lower Limbs.*
7. *Eczema of the Hands and Feet.*

Eczema Pilare.

Eczema Pilare attacks those parts of the body which are covered with hair, and on the whole is more difficult to treat than any of the other varieties. It commences with a mixture of vesicles and pustules; these in time burst and discharge a copious viscid secretion which dries and forms thick crusts. The hairs necessarily become involved in these crusts, and the result is a clotted mass of rapidly decomposing material which emits a peculiar faint sickly odour. When the crusts have been removed either by lotions or poultices, the subjacent skin is left raw and moist. Successive incrustations follow and often protract recovery to an indefinite period.

Eczema of the Face and Ears usually attacks corresponding parts on the two sides of the face. Not limiting itself to the actual skin, it is apt to spread to the eyes, nose, or mouth, giving rise to Eczematous inflammation in the mucous membrane of these organs. When it attacks the ear, it causes swelling and tension similar to that which occurs in Erysipelas of the part. The accompanying inflammation of the Meatus explains not only the deafness which often attends this form of Eczema, but also the presence within the ear of crusts and other products of secretion. A long time generally elapses before desquamation sets in, and then deafness may again result from scales, wax, and hairs forming a solid mass and choking up the auditory canal. <small>Eczema of Face and Ears.</small>

Eczema of the Breasts usually exists in connection with pregnancy, lactation, or the presence of the ACARUS SCABIEI. It is confined almost exclusively to women and often implicates the subcutaneous cellular tissue, forming little abscesses therein. <small>Eczema of the Breasts.</small>

Eczema of the Navel is generally very intractable both from the difficulty of keeping applications to the part and from the constant friction to which it is exposed. It often coincides with Eczema of the belly, and is important mainly for the following reason, that an inexperienced or unwary practitioner might mistake it for a syphilitic affection which frequently occurs in that region and which I shall subsequently notice. <small>Eczema of the Navel.</small>

Eczema of the Genitals is a most important variety and may <small>Eczema of the Genitals;</small>

be considered under two heads, according as its seat is in the male or female organs.

in the female, When the eruption attacks the female genital organs, it becomes one of the most intractable complaints with which the surgeon has to deal. The accompanying irritation is so intense that the patient cannot resist scratching, and the sensation of pleasure and relief caused thereby is frequently the origin of masturbation with its long train of concomitant miseries. When the disease has lasted for some time, it produces inflammation and hypertrophy of the mucous membrane with scropurulent secretion—in fact, the usual symptoms of gonorrhœal vaginitis -so that it frequently becomes a difficult matter to establish a diagnosis between the two affections. In the Eczematous form, however, it should be borne in mind that the perinæum and labia majora are pretty sure to be involved; that the discharge is less purulent; that the irritation is far more intense; and that the symptoms are more severe towards the catamenial period.

in the male, attacking (a) Penis In the male sex it attacks the penis, especially the glans and prepuce, and, owing to the flaccidity and sponginess of these parts, is generally accompanied with considerable œdema and infiltration. Fissures and excoriations are of frequent occurrence and are apt from the swelling of surrounding parts to present an appearance of great depth. The affection in this stage is liable to be confounded with a chancre, especially as it is wont to appear after repeated intercourse with a fresh female to whose secretions the male organ has not yet become accustomed. The points to be borne in mind in making a diagnosis between the two diseases are mainly as follows: (1) the early manifestation of the Eczematous lesion—usually a day or two after intercourse; (2) its speedy subsidence; for, if the parts be kept at rest, the excoriation heals up in the course of a few days and the œdema subsides; (3) the frequent coexistence of Eczema on some other part or parts of the body. Notwithstanding the readiness with which this disease perhaps yields to treatment at the time, relapses are very frequent, its tendency being to become chronic and most intractable.

Eczema, when it attacks the scrotum, is characterized by successive formations and desquamations of fine scales. The intense irritation in this, one of the most tiresome varieties of Eczema, must always be dreaded on the score not only of present suffering but also of possible consequences. The desire to scratch or rub the parts affected is overpowering; and its gratification, whilst affording slight momentary relief, never fails most seriously to aggravate the eruption. All this manipulation of the sexual organs is sure to be attended with more or less excitement of the sexual feelings; and it is in this manner that Eczematous eruptions in these parts not unfrequently give rise to the habit of self-abuse. On the subsidence of the local irritation the habit may perhaps be relinquished; but too often it is persisted in after the latter has ceased, and gradually gains a dominant influence over the patient. Then ensues that long train of consequences, sad enough to tell, sadder still to bear—loss of appetite, colour, flesh and strength; nervous despondency; incapacity for mental or physical exertion; and either sexual debility or complete impotence.

Eczema of the Perinæum, Groins, and Margin of Anus may occur in either sex. It furnishes a quantity of plastic secretion which stains and stiffens the linen and is a source of great annoyance to the patient. The cure in these cases is most tedious on account of the proneness of the eruption not merely to recur but also to spread beyond its original confines. When

(b) Scrotum.

Eczema of Perinæum and adjacent parts.

it attacks the anus, the intolerable irritation drives the patient to scratch or rub the part and so to aggravate the mischief. The consequences are excoriations and fissures, differing from ordinary fissures of the anus in their number, their superficiality, and their treatment—no operation being required, but merely the local and general employment of anti-dartrous remedies.

Eczema of Lower Limbs.
Eczema of the Lower Limbs usually coincides with and depends upon a varicose condition of the veins. In most cases of varicose veins in the lower extremities, ulcers will form on very slight provocation, extend rapidly and, unless treatment be directed to the primary cause, set medical skill at defiance. Eczema, when associated with an ulcer of this description, is cured with great difficulty and leaves behind it a blackish stain, due to increased deposit of pigment in the skin tissue.

Eczema of Hands and Feet
Eczema of the Hands and Feet presents different characters, according as it is acute or chronic.

(a) Chronic.
Chronic Eczema of the Hands is characterized by excoriations and fissures secreting a sero-plastic fluid which hardens and forms crusts; secondly, by a number of little papules similar to those met with in Lichen. It prevails among persons whose occupations oblige them to handle irritating liquids, powders, &c., as for instance, chemists, painters, grocers, confectioners, and cooks. Many persons daily expose their hands to all kinds of irritating substances without bad consequences; but those, in whom the Dartrous predisposition exists, cannot do so with the same impunity.

(b) Acute.
Acute Eczema of the Hands prevails generally towards the end of spring and during the height of summer. The parts attacked swell; as the swelling increases, the skin becomes red and causes a feeling of heat, irritation, and tension. Little

vesicles next appear which rapidly increase in size; the partitions between them break down and form by their coalescence larger vesicles, sometimes equal in size to the bullæ of Pemphigus. If the epidermis is too strong to break, the vesicles gradually fade away, leaving behind them a scaly irritable condition of skin. If, on the other hand, the vesicles break either spontaneously or by friction, excoriations ensue and crusts form which, after protecting the surface for a longer or shorter time, at length fall off and the disease is at an end.

When Eczema attacks the feet, it presents much the same appearances as in the hands; the main difference being, that the thickness and toughness of the skin of the feet usually prevent rupture of the vesicles.

Treatment of Eczema.

In the first stage of Eczema, when the vesicles are still intact and when a certain amount of inflammation is present, antiphlogistic measures must be adopted. Warm baths and emollient lotions—such as decoction of linseed or poppy-heads, thin starch and water, oatmeal gruel—will form the most appropriate local treatment. All applications which tend to break the vesicles must be strictly avoided, and every effort must be made to promote absorption of their contents and prevent suppuration and formation of fissures. *Acute Eczema. Treatment of the first stage, locally,*

Should the eruption take on an Impetiginous character, our object should be by means of poultices to accelerate the rupture of the pustules and vesico-pustules. The best sort of poultice for this purpose is rice-flour or potato-flour moistened with water; linseed-meal is apt at this stage to do more harm than good, on account of its proneness to ferment. Dry powders, such as the flours just mentioned, may also be tried with advantage.

internally. As a rule, no constitutional treatment is needed beyond cooling and refreshing drinks, such as lemonade, orangeade, barley-water, and restriction to a light and simple diet. If, however, there be any tendency to plethora, it is safer to give a brisk aperient and then to act on the skin and kidneys; for which purpose the following will be found an efficient combination:—

℞ Potassæ Bicarb. . ℈j
Potassæ Nitrat. . gr.x
Ant. Pot. Tart. . gr.¼
Mist. Amygd. ad ℥j

Misce ut fiat haustus—to be taken night and morning for two or three days.

Treatment of the second stage, In the second stage (i.e. the rupture of the vesicles and the appearance of the secretion), the internal treatment should consist of mild vegetable aperients, such as Manna, Castor-oil, *internally,* Rhubarb, Senna, &c. in small but frequent doses. Remedies of too drastic a nature must be carefully avoided. The Sulphates of Potash, Soda, Magnesia, and other salines, so popular among some practitioners, require such large doses to be of service that they are apt to weaken the intestines. The vegetable aperients above recommended may be advantageously combined, in weakly subjects, with mild vegetable tonics. I have found the following formulæ exceedingly useful in such cases:—

℞ Mannæ Opt. . . ʒij   Pulv. Rhæi . . ℈j
Sp. Ammon. Arom. . ʒss   Syr. Zingib. . . ʒj
Ext. Glycyrrhizæ . ℈j   Inf. Aurantii Comp. . } ℥iss
Inf. Sennæ Comp. . ℥iss   vel Inf. Gentian. Comp.
Misce ut fiat haustus.   Misce ut fiat haustus.

One or other to be taken fasting, every other morning, until the eruption begins to desiccate.

locally. Locally this stage requires the same treatment as the first, viz. emollient poultices and lotions, soothing-powders, &c. The baths, however, should be given at a somewhat higher temperature. When the head and face are the seats of eruption, vapour-baths prove of great service. But perhaps the most convenient and efficient of all are the baths of pulverized water—*Bains à Hydrofère*, as the French call them. In these the action of the continuous stream of finely pulverized water appears to be more beneficial than simple contact with water or vapour, facilitating the detachment of the crusts as well as allaying heat and irritation in the subjacent skin.

When the eruption has reached the third stage, the remedies just mentioned become inappropriate. The patient now needs one or more of the vegetable or mineral tonics such as Quinine, Lupuline, Gentian, Steel, &c. together with a more generous diet. Cod-liver oil is a drug of great service in this stage, especially among the young or ill-nourished. If after a trial of such remedies the eruption still lingers, recourse must be had to Arsenic—the great specific in Dartrous affections, more especially those of an Eczematous nature. Should the Dartrous affection exist alone, unmixed with any other diathesis, Arsenic will nearly always effect a cure. *When* and *how* to administer this useful but dangerous remedy requires great judgment and experience. In the first and second stages its exhibition is not only useless but injurious, since it stimulates the skin and so increases the inflammation; but when all inflammation has subsided and the eruption has reached the third stage, or when it has passed from the acute into the chronic form, then Arsenic becomes invaluable. It is administered far better in solution than in pills, as in making a quantity of the latter it is extremely difficult to allot to each its exact proportion of the mineral.

Treatment of the third stage, and of chronic Eczema.

Tonics.

Cod oil.

Arsenic.

Of the various preparations of Arsenic employed in this country and on the Continent, experience inclines me to give the preference to the so-called 'Fowler's Solution,' the *Liquor Arsenicalis* of the British Pharmacopœia. I usually commence with two minims, and gradually increase the dose up to five minims, three times a day; directing it to be taken during meals, and occasionally intermitting its use on account of its known tendency to accumulate in the system and so give rise to uncomfortable symptoms. Should the stomach refuse to tolerate the drug even in very small doses, I am in the habit of combining it with a little Laudanum; thus protected, it usually ceases to offend even an irritable stomach.

Sulphur. Sulphur stands next to Arsenic as a serviceable drug in the treatment of this stage of Eczema, and, like Arsenic, should be avoided in the treatment of the first and second stages. It can be given in powder or in electuary, or by the use of those mineral waters which contain it as a natural ingredient, such as the Spas of Enghien or Aix-la-Chapelle on the Continent, or of Harrowgate in our own country.

Cantharides. Cantharides (in the form of Tincture) is much extolled by MM. Devergie and Ricord as a specific in one of the most obstinate forms of Eczema, viz. Lichen Inveteratus. MM. Hardy and Bazin likewise maintain its *occasional* utility, provided certain precautions are observed in its administration —the precautions, namely, of giving it in small, gradually increasing doses, and suspending its use immediately on the appearance of symptoms indicating irritation of the urinary apparatus.

Other remedies. Innumerable are the remedies which from time to time have been extolled as specifics in Eczematous eruptions. To enumerate them and their imputed virtues would be mere waste of time.

There are, however, some of these, concerning whose indiscriminate use I will make a few remarks. The sudorific woods, such as Sassafras, Guaiacum, and Sarsaparilla, are not only useless but positively injurious during the acute stage of the eruption. Again, Iodide of Potassium is too often exhibited in a haphazard manner, to the aggravation of the skin-disease as well as the detriment of the patient's health. Invaluable as are the different preparations of Iodine in many other kinds of skin-disease, experience has certainly proved them to be of little or no service in Eczema. Some time ago, certain French Dermatologists blazoned forth the transcendent virtues of the Hydrocotyle Asiatica in Eczema, Lepra and Psoriasis. From a drug possessing such a sesquipedalian high-sounding title great things were expected, but were not forthcoming. It proved utterly inert and its use was soon relinquished.

In chronic Eczema certain local applications may be advantageously tried, as for instance the following ointment:— *Local treatment. Ointments.*

℞ Bismuthi Nitrat. . ℨ ij ⎫
  vel Hydrarg. Chlorid. . ℈ j ⎬ Misce.
  Ol. Amygd. Dulc. . ℨ ij ⎪
  Adipis . . . ℨ vj ⎭

In Lichen Inveteratus Oil of Cade, mixed with cold cream or glycerine and starch, has been found very successful. In cases of Pityriasis succeeding Eczema, the use of a mild sulphur ointment often accelerates the cure. When the disease attacks the external genital organs, local applications are indispensable both for allaying irritation and forestalling those evils to which I before made allusion (p. 23). I have usually found the following an efficient combination for this purpose:—

℞ Potassii Cyanid. . . gr. vj ⎫
  vel Zinci Oxyd. . ℨ j ⎬ Misce.
  Ol. Amygd. Dulc. . ℨ ij ⎪
  Unguent. Cetacei . ℨ v ⎭

When the inflammation has invaded the urethra, vagina, or anus, the occasional introduction of a catheter, bougie, or small speculum, coated with some emollient astringent preparation, is often of service both in allaying irritation, soothing the patient, and restoring fresh tone and vigour to the relaxed and enfeebled membranes. I frequently find the following formula answer well:—

℞ Zinci Oxyd. . . ʒj—ʒij ⎫
   Camphor. Pulv. . ʒss—ʒj ⎪
   Sp. Vini Rect. . . ʒj   ⎬ Misce.
   Ol. Amygd. Dulc.,   ⎪
   Unguent. Cetacei . āā ʒij ⎭

Of course delicate and careful manipulation is needed in employing these useful but, in unskilled hands, dangerous instruments. As to specula and bougies, I am fully aware of the disfavour, nay, the pious disgust (real or assumed), with which the use of these instruments is regarded by a certain section of the Profession. The speculum especially I have heard stigmatized as an outrage upon feminine modesty and refinement, or else as a pander to the sickly prurient longings either of the patient or her Doctor or both. That it is sometimes abused I do not for a moment deny; but its abuse is one thing, its use is another; and I have yet to learn why an instrument, invaluable for the detection and treatment of certain diseases, is to be discarded simply because it happens to be employed in an objectionable manner by a certain class of practitioners.

Counter-irritation.
In very obstinate indolent cases, counter-irritation by means of one of the ordinary vesicants is sometimes serviceable, partly because it sets up a new action in the capillaries of the part, and partly because the sharp pain of the blister is found more endurable than the chronic worry of the disease. In such cases local treatment may be aided by the internal exhibition of sedatives, as for instance Opium, Belladonna, Hyoscyamus, &c.

Another class of local remedies are baths, which should be merely emollient in the two first stages, but more or less stimulating in the third. In the dry and papular varieties of Eczema alkaline baths, carefully regulated so as gently to stimulate without irritating the skin, allay inflammation and hasten the cure. Sulphur-baths should contain but a small proportion of the active ingredient, and should, as a rule, be reserved for the dry desquamating period of Eczematous eruptions. *Baths.*

When Eczema proves very intractable, a course of some of the natural mineral waters is often attended with advantage. M. Hardy especially recommends the waters of St. Gervais in Savoy. This Spa seems to exercise a sedative influence peculiarly beneficial to nervous, excitable, and irritable patients. In Lichenous varieties of Eczema, where the skin has become greatly altered in thickness and texture, the mineral waters of Leuk in Switzerland often produce most favourable results. Sea-bathing is rarely tolerated in Eczema, its tendency being to aggravate and prolong the mischief. *Mineral Waters.*

Last but not least, hygiene has to be considered; and a very powerful ally it may prove on the surgeon's side, both in promoting the cure of the affection and in preventing its recurrence. Rigid abstinence from highly-seasoned dishes, stimulating drinks, shell-fish either *au naturel* or in salad, acid fruits and vegetables, is absolutely necessary. Due avoidance of all excess either at or after table and of over-indulgence of the passions and emotions, together with relinquishment of occupations and trades predisposing to cutaneous diseases, will greatly assist the surgeon in improving his patient's condition and fortifying him against those frequent relapses which are so troublesome a feature in this complaint. *General Hygiene.*

PITYRIASIS.

Pityriasis: its general characters,

Having now concluded the subject of Eczema and its numerous varieties, I will proceed to consider the remaining Dartrous eruptions; viz. Pityriasis and Psoriasis. These are characterized by special lesions and by the absence at every stage of that humid secretion which formed so marked a feature in Eczema. The former presents a slight inflammation of the superficial layers of the skin, accompanied by exfoliation and constant succession of little dry furfuraceous scales, in which respect as well as in the absence of all moist secretion it differs from the Pityriasis which is merely the final stage of Eczema.

and several varieties.

The form, appearance, and situation of the disease suggest three varieties; viz. *Pityriasis Rubra, Circinata,* and *Pilaris*.

Pityriasis Rubra.

Pityriasis Rubra commences by well-defined red stains, and is always accompanied with some slight amount of heat and irritation. It attacks chiefly the face, neck, chest, feet, and hands; and, if it do not become chronic, yields to treatment in from three to eight weeks. A form of Pityriasis exists called 'Pityriasis Versicolor,' which might be mistaken for the affection just described; but the microscope, by demonstrating the presence of the parasite in the former, will always ensure the diagnosis between the two.

Pityriasis Circinata.

Pityriasis Circinata differs mainly from the preceding variety in the number of rose-coloured disks or segments of circles covered with fine scales—these latter being well-marked at the commencement, but less distinct towards the close of the disease.

In *Pityriasis Pilaris* the skin has a dry wrinkled uneven appearance in consequence of the scales choking up the orifices of the hair-follicles and thus forming little prominences. So obstinate a disease is this that it may almost be reckoned among the incurable alterations of the epidermis, such as Ichthyosis, Pellagra, &c.

Pityriasis Pilaris.

Treatment of Pityriasis.

In P. Circinata the internal use of Arsenic proves of great service both in aiding the cure and preventing relapses. Externally much benefit is often derived from the *Unguent. Creasoti* of the British, or the *Unguent. Picis Liquidæ* of the old London Pharmacopœia. If these remedies fail, mild baths of Sulphur vapour will be worth trial.

Treatment of each variety. (a) medicinal,

In the other two varieties Oil of Cade combined with glycerine or starch-paste; ointments composed of the salts or oxyde of Mercury; Sulphur baths, and baths of pulverized water (see p. 27), are among the most successful remedies. A visit to the Spas of Harrowgate, Aix-la-Chapelle, Bareges, or some of the other natural Sulphur springs, will occasionally effect a cure when the disease has resisted all other measures.

Of course in Pityriasis as in all other Dartrous affections, hygienic and dietetic principles are of the greatest importance. Good air, large rooms, simple food, abstinence from alcoholic drinks, early hours, and avoidance of undue physical exertion and mental excitement, are indispensable conditions in carrying out any line of treatment to a successful issue.

(b.) hygienic.

PSORIASIS.

Psoriasis was evidently considered by the old Dermatologists to be identical with Lepra; in fact, under this latter term they included a variety of diseases now acknowledged to be generically distinct. Though liable to attack all parts of the tegumentary surface, it has a predilection for the elbows and knees, the outer surface of the limbs, the skin of the scalp and chin. It commences by an eruption of patches of thick white shining scales, composed of successive accumulations of epithelial lamellæ, dry and closely adherent to the skin. The appearance of these patches may be best described by comparing them to the droppings of a wax-candle. The subjacent surface is red, uneven, and thickened and, when the eruption occupies the flexure of a limb, more or less deeply fissured. There is usually a certain amount of irritation, which increases towards night and whenever from any cause the circulation becomes quickened. The patient at first may enjoy perfect health, but, as the disorder becomes more developed, he rarely fails to suffer some amount of visceral disturbance, such as impaired appetite, irritability of stomach, diarrhœa, &c.

Psoriasis is essentially a chronic eruption. Judicious treatment and careful regimen may seemingly cure the disease for a time; but any little excess at table or over-exertion, whether of mind or body, is apt to bring it back with renewed vigour. It is one of the most baffling diseases which fall to the

Its favourite seats;

General characters;

surgeon's lot. In spite of the most judicious and persevering treatment, it will often cling to a patient for months and years, sometimes even his whole lifetime.

The varieties of Psoriasis recognised by different Dermatologists are endless. Thus writers describe a *Psoriasis Guttata*, *P. Gyrata*, *P. Diffusa*, *P. Circinata* (common *Lepra*), *P. Inveterata*, and so on. But these are after all mere modifications of one and the same disease, the salient peculiarity of each being indicated by its adjectival name, and therefore they do not need a separate description. Other varieties, however, exist which have their seat in particular parts or organs of the body; and these give rise to so much discomfort, disfigurement, or both, that they call for more than mere enumeration of their names.

Psoriasis Capitis. This form commences nearly always at the front of the scalp, whence it extends to the forehead, back, or sides of the head. It does not commonly affect the hairbulbs; hence the hair not having its nutrition interfered with rarely falls off, and, if it does, soon grows again after the subsidence of the eruption. Psoriasis is rarely confined to the head alone; more often it makes its appearance simultaneously on some other part or parts of the body.

Psoriasis Unguinum is characterized by deep fissures in the nails, giving them a coarse rough irregular wrinkled appearance. The nail often falls off and is replaced by a scaly crust, which in its turn becomes detached to be succeeded by a natural nail when a healthy action has been set up in the matrix.

Psoriasis Palmaris et Plantaris. This form may either be limited to the palms of the hands or soles of the feet, or may occupy the entire surface of the extremities. In consequence of the large thick scales, the deep fissures, and the thickness of the skin, acts of prehension and locomotion are performed imperfectly and with inconvenience. Psoriasis, when confined to the hands and feet, points almost invariably to a Syphilitic taint of the system. A case, which was under my observation a few months ago, furnished a striking illustration of this fact; the man had had Syphilis, and the several extremities of his body, viz., the hands, feet, nose, and prepuce, were covered with scales of Psoriasis.

Psoriasis Faciei is a rare affection and lucky it is so, for the unfortunate patient has no means of concealing his disfigurement from the eyes of his fellow-creatures. When it attacks the eyelids, it becomes very intractable and is apt to entail severe consequences. The increased thickness of skin and the existence of scales, however small, impede the action of the eyelids and in aggravated cases may even induce Entropion or Ectropion. The extension of the disease to the Conjunctiva sets up chronic inflammation of that membrane.

Psoriasis Præputialis. This variety of Psoriasis may involve the penis with its glans, prepuce, and scrotum. It presents soft delicate scales separated by deep painful moist fissures, which render erection very painful and intercourse almost impossible. The great and sudden differences in size to which the male organ is subject, so incompatible with an indurated fissured condition of its skin, tend to aggravate and prolong this painful disease.

Of all temperaments the sanguine seems most predisposed to the developement of Psoriasis. It occurs more commonly in the male than the female sex, and attacks subjects in the very prime and full vigour of manhood. All those causes which tend to arouse the Dartrous diathesis have an influence in evoking Psoriasis, as for instance errors in diet, abuse of alcoholic drinks, dissipation, depressing emotions, overwork of body or mind, &c.

TREATMENT OF PSORIASIS.

In Psoriasis as in all other Dartrous affections, Arsenic is the remedy on which most reliance is be placed. Here, as in Eczema, it must not be exhibited until the acute or inflammatory stage of the eruption has passed away and the epithelial scales have commenced falling off. Previous to this, tepid baths, soothing local applications, and mild aperients, are all that is requisite. Physicians and surgeons have in all ages vaunted numerous drugs as specifics in this disease, but, after giving them all a fair trial, experience has fully shown that Arsenic is the only efficient medicine. Indeed, Arsenic in Psoriasis seems to have the same effect in accelerating the cure and preventing relapses, as Mercury has in Secondary Syphilis. It may be exhibited in various forms, but there are none safer and more efficient than the *Liquor Arsenicalis* of the British Pharmacopœia. I would here repeat the precautions which I before mentioned when discussing the treatment of Eczema (pp. 27-28) viz., that the Arsenic should be taken during a meal and in small gradually-increasing doses; that it should be intermitted from time to time, and always suspended immediately on the appearance of any symptom of its affecting the system injuriously. Internal treatment. Arsenic.

The next most valuable remedy in Psoriasis is Cantharides, which, on account of its powerful influence on the genito-urinary apparatus, must be employed with great judgement and caution. Its most convenient preparation is the Tincture. The dose should begin with three or four minims twice a day in any demulcent fluid, and be increased one minim daily until it reaches half a drachm. Cantharides.

Copaiba.

Copaiba has occasionally been found beneficial, but neither it nor Cantharides produce the same well-marked beneficial effects as Arsenic. The nauseous eructations and gastro-intestinal disturbance, which it so frequently occasions, are great drawbacks to its use. It may be conveniently given in the form of pills with Carbonate of Magnesia as the excipient and a very small quantity of Opium to guard its action on the intestinal canal. These pills, if coated with silver or gum, possess neither smell nor taste and will rarely offend even a delicate stomach.

External treatment.

The local therapeutic measures comprise baths and ointments, and these will sometimes cure Psoriasis without the assistance of internal treatment; usually, however, their conjoint influence is necessary.

Baths.

The baths should at first be simply emollient. At a later period, when Arsenic is being exhibited, they should be slightly impregnated with Sulphur so as to stimulate the skin. Vapour-baths, alkaline baths, and pulverized-water baths, are also of great service, and will suggest themselves according as the surgeon sees indications for their employment. It must not, however, be forgotten that in this as in all other forms of skin-disease they may prove remedies powerful for evil as well as for good. All depends upon their being employed in the right cases, in the right way, and at the right time: consequently, they should never be taken except under medical direction, and with the assistance of skilled and careful personal attendants. Mineral baths may be taken either at home, artificially prepared, or, should the patient have time, money, and leisure for the trip, he may combine pleasure with profit by a visit to some of the sulphurous or alkaline Spas in England or on the Continent.

Ointments.

The ointments most efficacious in the treatment of Psoriasis are those which contain Sulphur, Mercurials, Tar, or Oil of Cade, in small proportions. I usually employ the *Unguentum Sulph.* the *Ung. Hydrarg. Nitratis,* or *Nitrico-Oxyd.* of the British, or the *Ung. Picis Liquidæ* of the old London Pharmacopœia, diluting each with olive or almond oil to such strength as the case may require. Equal parts of the Sulphur and Pitch ointments just mentioned, more or less diluted, form a most useful application in cases of inveterate Psoriasis. For applying Oil of Cade the following formula, invented by M. Hardy, will be found superior to any other:—

With one ounce of pure glycerine previously warmed mix starch in sufficient quantity to form a semi-fluid paste. Then add very gradually a drachm or a drachm and a half of Oil of Cade, rubbing it down with the paste in a mortar, until the several ingredients are thoroughly mixed. After cooling, the ointment is fit for use.

They are chiefly the chronic obstinate cases in which, according to my experience, pitch, tar, and remedies of this kind prove serviceable.

Strict attention to hygiene plays an important part in the treatment of this, as of all other skin diseases. Plain food, temperance in the use of alcoholic or stimulating drinks, early hours, avoidance of heated rooms and scenes of dissipation, regular work and domestic life—are indispensable conditions for a cure. Secret neglect of such hygienic precautions on the part of the patient is too often the real explanation of those cases, which seem to defy all treatment and bring unmerited censure on the medical attendant. Experience daily proves that, troublesome and difficult as it may be to treat the disease, it is often much more troublesome and more difficult to treat the patient. Professing compliance with his Doctor's orders, he none the less turns from his evil habits; and, when at length he has to pay the penalty of his disobedience, he lays the blame on every one but the real culprit.

General Hygiene.

ARTHRITIC ERUPTIONS.

Arthritis: its general characters, Arthritis, as defined by M. Bazin, is a constitutional disease, characterized by a tendency to the formation of a peculiar morbid product, and by various affections of the skin, joints, and viscera, which usually terminate in resolution. It occurs principally in gouty and rheumatic subjects and, according to the authority just quoted, is accompanied by the following symptoms: disordered action of the skin with increased perspiration, especially in the head, arm-pits, hands, feet, and sexual organs; premature baldness; tendency to obesity in spite of a moderate appetite; habitual constipation; headache; a carious condition of the teeth; varicose veins and hæmorrhoids; pains in joints and limbs; and great sensibility to cold and changes of temperature.

and different stages. The first stage is marked by slight superficial temporary affections of the cutaneous or mucous membrane, such as coryzas, sore throat, bronchitis, and certain pseudo-exanthematous eruptions, e. g. Zona, Urticaria, and Furunculus. The cutaneous manifestations usually appear in spring and summer, the catarrhal in autumn and winter.

The second stage presents tegumentary affections of a more obstinate and localized character than those of the preceding stage, alternating or coinciding with acute attacks of gout and rheumatism. The more intense the latter, the less pronounced the former, and *vice versâ*. Prurigo of the anus and genitals is the most troublesome eruption of this stage.

OF CONSTITUTIONAL ORIGIN.

The third stage is characterized by more or less crippling and disorganization of the joints from Arthritic inflammation and the resulting 'tophaceous' deposits. The general health fails and the patient falls a victim to disease of some internal organ, usually the brain, heart, or kidneys.

So much for the general characters of the Arthritic *Diathesis*; I will now briefly state what M. Bazin considers to be those of Arthritic *Eruptions*.

They are non-contagious, seldom attack children, and are almost always evoked by meteorological influences. They resemble the Dartrous eruptions in their mobile character and in their mode of progression from the cutaneous and mucous to the serous and synovial membranes, but have no tendency to destroy or perforate the tissues, as have Syphilitic and Scrofulous eruptions. *[margin: Arthritic Eruptions. Their general characters]*

Whatever the seat of the affection, pain more or less intense is an accompaniment, but of quite a different character to the irritation present in Dartrous affections. *[margin: as to Pain;]*

They generally attack the uncovered parts of the body and those richly supplied with sweat-glands and hair-follicles, e.g. the face, forehead, scalp, neck, chest, hands, feet, arms, sexual organs, the axillary and umbilical regions, and the breasts at the time of lactation. *[margin: Seat;]*

They are usually circular and well defined, appearing at first in perfectly distinct patches, which by their subsequent union may cover a large surface. Having no propensity to attack fresh surfaces, they never become general like Dartrous eruptions. *[margin: Form;]*

The colour is that of red wine or cherry juice and is caused by a varicose condition of the capillaries, which become so dilated as sometimes to burst and form little punctated spots of hæmorrhage. *[margin: Colour;]*

Secretion; The affected parts present great deficiency, sometimes total absence, of secretion; and are usually covered either with scales or thin crusts.

Complexity; In no other class of eruption is there so large a mixture of elementary lesions. Thus Lichen, Pityriasis, and Eczema, with their papules, scales, vesicles, and pustules, may—and often do—make their appearance simultaneously on the skin of an Arthritic patient, but almost invariably in an unsymmetrical manner, as for instance on one arm, one hand, one leg, &c.

Progress; The Arthritides are far more tenacious in the early than in the later stages of their progress; and, when a relapse occurs, the eruption usually returns to the same spot as before. In both these respects, no less than in their want of symmetry, they present well-marked points of difference from Dartrous eruptions—the latter being symmetrical; more tenacious, the longer they last; and having a strong tendency, in cases of relapse, to invade fresh ground and thus gradually to traverse the whole cutaneous surface.

Behaviour with remedies. In Arthritic eruptions the anti-dartrous remedies (Arsenic, Sulphur, &c.) are, according to M. Bazin, utterly useless; the Alkalies, Colchicum, and other anti-rheumatic remedies, alone being of any service.

VARIETIES OF ARTHRITIC ERUPTIONS.

Varieties of Arthritic eruptions recognised by M. Bazin. To give a full description of each variety of Arthritic eruption, as recognised by M. Bazin, would occupy far more space than can here be afforded. I shall therefore content myself with giving merely a brief outline of the groups into which he divides this class of skin diseases, entering into detail only in the case of the more common and therefore more important

OF CONSTITUTIONAL ORIGIN. 43

varieties. For the sake of clearness I will give M. Bazin's classification in a tabular form.

M. BAZIN'S CLASSIFICATION OF THE ARTHRITIDES.

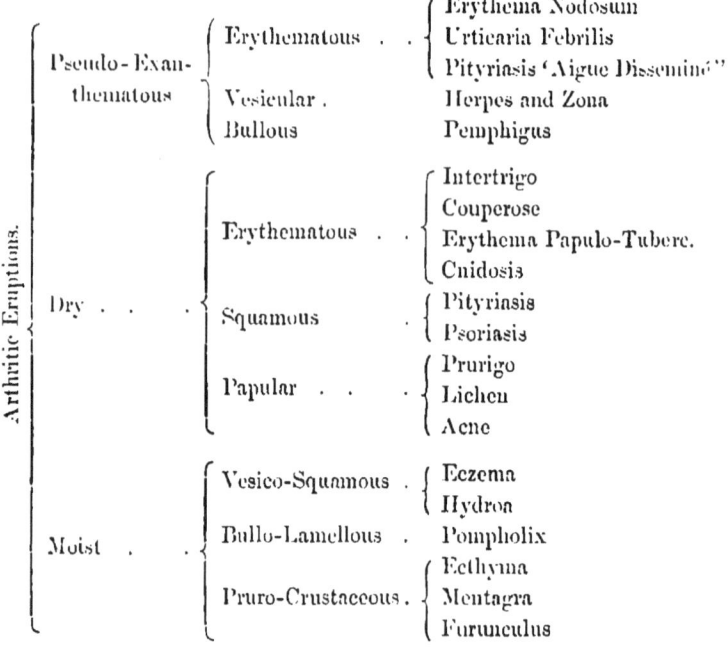

I will now describe a few of the most important of these Arthritic affections. Of the *Pseudo-Exanthematous* the following call for notice:— (A) The Pseudo-Exanthematous.

Erythema Nodosum is ushered in by general malaise and febrile disturbance, followed by an eruption of red hard painful patches, raised in the centre, indurated at their base, and disappearing in from twelve to fifteen days. There is usually much irritability in the parts about to be attacked, as well as pain in the joints and limbs. Erythema Nodosum.

Urticaria Febrilis, the common 'nettle-rash,' commences with headache, lassitude, gastric disturbance, and febrile symptoms. Urticaria Febrilis. Cnidosis.

Locally the first symptom is a pruriginous sensation, followed by the appearance of a number of rosy papular patches, which are generally circular, hard towards their outer border, and evanescent on pressure. The papules may last from a few minutes to a few hours and are soon replaced by others, the whole period of the attack occupying from seven to ten days. In this and the last affection, a restricted diet, cooling drinks, mild aperients, fomentations locally and alkalies internally, form the appropriate treatment. *Cnidosis* is merely a variety of Urticaria.

Herpes and Zona. *Herpes* usually begins with general malaise, loss of appetite, febrile disturbance, heat and irritation, followed by the appearance of a few red spots on some part of the skin. These spots are soon covered by a number of vesicles of various sizes, hard, globular, and transparent, each being encircled by a distinct red areola, so that an uniform red tint is imparted to the whole surface of the group. Towards the third day the vesicles are replaced by small brown or yellow crusts; these after a while fall off, leaving behind them red stains and sometimes rather painful excoriations. All parts of the body are liable to this affection, but more especially the lips, cheeks, neck, chest, prepuce, labia majora and minora. Its usual exciting cause is some excess or imprudence in diet.

When Herpes is disposed in the form of an oblique flexuous semicircle, starting from one point of the median line of the body and terminating at the opposite point, it constitutes that well-known painful affection which Dermatologists call *Zona* and the vulgar call *Shingles.*

Pemphigus. *Pemphigus* is ushered in by febrile symptoms, lassitude, local irritation and pain, followed by the appearance of red circular patches, the epidermis of which is soon raised by a copious serous exudation forming bullæ of various sizes. These bullæ are after a while replaced by little brown foliaceous crusts, often associated with boils and Ecthyma. For some time after the detachment of the crusts, the subjacent skin is usually stained more or less of a violet colour.

Pemphigus, Herpes, and *Zona* require in the main the same treatment, viz. a light diet, acidulated or slightly diuretic drinks, mild aperients, and emollient or gently astringent local applications, such as Starch-powder, Glycerine, Oxyde of Zinc, &c.

Of the *Dry* Arthritides, the following chiefly claim consideration:— *(n) The Dry.*

Intertrigo. This affection is generally induced by want of cleanliness or by the irritation of unhealthy perspiration, and is characterized at the outset by an erythematous blush of the skin and much irritation. The inflammation increases in degree and extent, producing pustules, tubercles, and boils. The disease affects those parts of the body where two surfaces of the skin are in contact, e.g. the armpits, the fold behind the ears, the folds of the abdomen and groin in corpulent subjects, the anus and genitals. *Intertrigo.*

Strict attention to personal cleanliness, especially in those parts of the body where dirt is most likely to accumulate, as for instance the flexures of the trunk and limbs; the alternate use of alkaline and emollient baths; soothing powders, such as Rice-flour or Potato-flour, Oxyde of Zinc, Oxyde of Bismuth, &c.; these constitute the local treatment. Alkalies, either artificially prepared or as they exist in the natural mineral waters, are the most appropriate internal remedies.

Couperose is characterized by a distended varicose condition of the capillaries of the skin. The cheeks, nose, and labial *Couperose.*

commissures, are usually the parts attacked; but sometimes it selects the front of the chest towards the upper border of the Sternum. The colour deepens, and the skin becomes hypertrophied and covered with little red indurations which terminate in a pustule of the Acne kind. The disease is exceedingly obstinate and chronic, and throughout the whole of its progress is attended with heat, irritation, and shooting pains. Among its best ascertained causes may be mentioned excesses in food and alcoholic drinks; the use of irritating cosmetics, lotions, and powders; exposure of the face to excessive heat; and occupations which, by necessitating a stooping posture of the head, maintain a congested state of its cutaneous capillaries.

Pityriasis. Psoriasis. Lichen. The Arthritic forms of *Pityriasis, Psoriasis, Lichen* and *Prurigo*, do not differ from the Dartrous, except that they are accompanied with shooting stabbing pains and more or less evidence of a rheumatic diathesis. The constant sour perspiration, peculiar to Arthritic subjects, must be reckoned among the accidental causes of these affections.

Acne. Acne is an affection of the sebaceous glands, consisting in obliteration of the external orifice and dilatation of the canal from accumulation of the gland-secretion. M. Bazin recognises two forms, Acne *Hypertrophique* and Acne *Inflammatoire*. But, as Acne occurs more frequently and with more strongly marked characters in Scrofulous than in Arthritic subjects, I shall defer my description of this eruption until I come to speak of it under the head of Scrofula. (See pp. 96–99.)

(c) The Moist Eczema. Of the *moist* Arthritides I shall notice only three.

The Arthritic *Eczema* so closely resembles the Dartrous Eczema in all its main features, and the latter have been so fully described at the commencement of this work, that it is needless to enumerate them over again (see pp. 9–31).]

shall therefore merely explain those minor peculiarities, which, according to M. Bazin, the disease presents under the Arthritic form. They are: (1.) A predilection for the forehead, lips (especially the upper), nape of neck, temples; backs of fore-arms, hands and feet; anterior surface of legs; mammæ; and genital organs. (2.) Absence of symmetry, i. e. of a tendency to attack corresponding parts on the two sides of the body. (3.) Limitation of the eruption to small circumscribed patches of skin and disinclination to spread. (4.) Presence of pain or other symptoms indicative of rheumatic taint.

Mentagra is characterized by inflammation of the hair-folli- Mentagra. cles and may appear on any part where hair grows, but its favourite seat is the upper lip. It manifests itself in the form of a closely grouped mass of papules and pustules, yellow at summit, red and indurated at base. Boils, almost continual coryzæ, and painful fissures at the entrance of the nasal fossæ, are common and troublesome accompaniments of this disease. It is very obstinate and prone to recur; it attacks middle life and old age during the autumn and winter seasons; and usually the male rather than the female sex, the hair-follicles in the former being more strongly developed.

Epilation and the employment of the Oil of Cade, combined with the internal exhibition of alkalies, are the most successful remedies. Epilation removes the hair which seems to keep the inflamed follicles in a state of constant irritation; Oil of Cade alters the action of the part; while the alkalies counteract the constitutional mischief of which the Mentagra is only a symptom. Should the inflammation become severe, emollient poultices and soothing lotions must be applied. In recent cases epilation may be omitted, and an ointment of Glycerine with Red Precipitate substituted. Vapour baths, vapour or alkaline douches directed on the affected part, are useful adjuvants in the treatment of this affection.

Ecthyma. *Ecthyma*, with its allied form *Furunculus*, is an affection characterized by pustules with hard inflamed bases and summits covered with brown crusts. No part of the body is exempt from its attack, the limbs and neck being its favourite seats. It commences as a red elevation, the centre of which is occupied by a large vesicle filled with transparent serum. On or about the third day the serum becomes lactescent, and the centre of the vesicle depressed and marked with a black point. On raising the epidermis a little pus escapes and reveals a false membrane adherent to the dermis underneath it. In three or four days the epidermis breaks and forms with the false membrane a brown crust which falls off after a time, leaving behind it a violet scar with a central depression.

General Treatment of Arthritic Eruptions.

Having described the principal Arthritic eruptions and the special treatment needed by particular varieties, I will now enter more fully into the general treatment of this class of skin diseases. Such treatment must have reference to their prevention as well as to their cure.

Preventive treatment. Preventive treatment consists in removing as far as possible the known causes of the disease, and in a judicious application of hygienic principles. Patients predisposed to Arthritic affections should take every precaution to protect themselves against changes of temperature, should always wear flannel or woollen under-clothing from head to foot, and should frequently change the same because it is so apt to become saturated with perspiration. Professions or trades involving the manipulation of irritating substances should be relinquished. A simple unstimulating diet, consisting mainly of white meat and fresh vegetables, should be recommended, while dark meats, such as game and rich or spiced made-dishes, should be rigidly proscribed.

In Arthritic affections the internal and external use of alkalies is the most successful and at the same time most rational mode of treatment. Baths, composed of three ounces of Bicarbonate of Soda to about thirty gallons of warm water, form the best means of obtaining the influence of the remedy externally. Some surgeons, forgetful of the irritability of the skin in Arthritic subjects, prescribe baths far more strongly alkaline, and by so doing merely aggravate the disorder. The alkali most appropriate for internal use is the Bicarbonate of Potash; which should be taken in doses of from ten to twenty grains, dissolved in half a tumbler of water, two or three times a day. It is preferable to the Bicarbonate of Soda, because it is this latter alkali which, in combination with Uric Acid, forms the chief ingredient in the concretions peculiar to this diathesis. Free dilution with water accelerates the action of the remedy and facilitates the elimination of effete or morbid materials by the excreting organs.

Curative treatment.
(1) Internally.
Alkalies.

The natural alkaline waters of Vichy, Carlsbad, Ems, and Wiesbaden, are of great service in removing cutaneous affections of Arthritic origin. Sulphur waters are useful only when the Arthritic is combined with the Scrofulous diathesis, or when the latter has preceded the former. When the disease is nearly well, the Spas of Aix-la-Chapelle, Néris, and Bagnères de Bigorre, are useful in accelerating complete recovery.

Alkaline Spas.

Next to alkalies, Colchicum is the most valuable remedy in the treatment of these affections. It may be given alone or in

Colchicum.

combination with alkalies and other remedies which, by increasing the activity of the secretions, tend to purify the blood. The following will be found efficient combinations of the kind:—

 ℞ Tinct. Colchici Sem. . . ♏x—xv
  Sodæ Potassio-Tart. . . ʒij
  Ext. Taraxaci Fluid. . . ʒj
  Aq. Menth. Pip. . ad ʒj
  Misce ut fiat haustus bis die sumendus.

or,

 ℞ Ext. Colchici Acet. . . gr. j.
  Ext. Aloës Aquosi . . gr. iij.
  Pil. Plummeri . . . gr. v.
  Misce et divide in pilulas ij, omni nocte sumend.

It is almost needless to observe that Colchicum in many cases acts as a powerful depressant, and therefore should be taken only under the constant supervision of the medical attendant.

(2) Locally. The local treatment must of course vary with the character and stage of the eruption. It is impossible to say what are the precise remedies to be used in each particular case. As a general rule, however, it will be found that, so long as there is any moisture about the eruption, benefit will be derived from the application of simple emollient remedies, such as starch, rice-flour, or potato-flour. When on the other hand the eruption has become, or has been from the beginning, perfectly dry, some stimulating application such as Oil of Cade, either alone or made into a paste with starch and glycerine, will generally prove serviceable. Benzine may be substituted for Oil of Cade. It possesses the same virtues without the drawback of staining the skin.

Emollient Powders.

Oil of Cade.

Benzine.

 In Arthritic affections attended with great irritation, as so frequently occurs when the anal or genital regions are attacked, M. Bazin prescribes the following emollient formulæ:—

M. Bazin's Formulæ.

OF CONSTITUTIONAL ORIGIN.

Lotion

℞ Glycerine . . . . deux cuillerées.
Eau de Guimauve ⎱
ou Eau de Son ⎰ . . 500 grammes.

Liniment

℞ Eau de Chaux
Glycerine . . . . ā͞ā 30 gram.
Huile d'Amande Douce . . 60 gram.

Many other preparations, such as those recommended in the treatment of Eczema (pp. 25–31), tend to allay the local irritation when the disease is seated in the parts just mentioned. M. Bazin recommends Morphia as producing a counter-irrita- Morphia.tion in the skin far more endurable than the incessant and intolerable worry of the eruption. He prescribes an ointment composed of

℞ Morphine . . . 5 à 10 centigrammes
Axonge . . . . 30 grammes.

and directs a small quantity to be occasionally rubbed into the part affected.

SYPHILITIC ERUPTIONS.

To give a satisfactory account of Syphilitic skin-diseases, I must first of all consider that constitutional state in which they originate. In the Syphilitic ladder the skin disease is far from being the lowest round. I must therefore start from the bottom of the ladder and gradually work my way up to that point in the progress of Syphilis where the skin disease first makes its appearance.

In the olden time the various kinds of venereal affections were jumbled together in truly Babelish confusion. Medical men affiliated them all to one common parent, Syphilis. In the present day, however, *nous avons changé tout cela* and have deprived Syphilis of two of her children, Gonorrhœa and Soft Chancre. Voltaire quaintly and truly observed that 'Syphilis resembles the fine arts; no one knows the inventor.' However obscure the origin of Syphilis, it is now generally admitted that there are three distinct kinds of primary venereal affection; viz.,

Three distinct kinds of primary venereal disease.

(1.) Gonorrhœa; mentioned in Leviticus, chapter xv.

(2.) Soft Chancre; described by Celsus, Vella, Lecocq, and others; accompanied with chancrous bubo, but not followed by secondary symptoms.

(3.) Hard Chancre, with its series of secondary symptoms: a disease first noticed towards the close of the fifteenth century.

OF CONSTITUTIONAL ORIGIN.

Now, before describing Syphilitic eruptions, it is necessary that I should give a brief account of each of these three affections.

PRIMARY VENEREAL AFFECTIONS.

A. GONORRHŒA.

Gonorrhœa can be referred to the most remote ages as one of the consequences of unclean intercourse. Moses, Herodotus, Hippocrates, Celsus, Galen, and other ancient writers, each described affections identical with gonorrhœa and its complications; and it was not until the fifteenth century, when the appearance of a quasi-novel malady attracted universal attention by the intensity and persistence of its ravages, that gonorrhœa sank into comparative obscurity. The exact nature of this new disorder was warmly discussed; indeed it soon became the great bone of contention between the medical savants of the day. Fallope, Hoffman, Astruc, Hunter and others, considered gonorrhœa to be a symptom of Syphilis; Balfour, Duncan, Bell and others, made it a distinct affection. In the present day, medical authorities are almost unanimous as to the non-identity of the two affections. *History of Gonorrhœa.*

Gonorrhœa is characterized by the discharge of a muco-purulent fluid from the urethra, with inflammation usually of an acute kind, and more or less pain on passing water. The inflammation generally commences in the fossa navicularis, whence it invades the whole length of the urethra and sometimes the prepuce, glans, and submucous cellular tissue, causing balanitis, excoriations, vegetations, together with inflammation of the prostate, epididymis, and neck of the bladder. *Symptoms.*

The exciting cause is almost invariably intercourse with a *Cause.*

female who is herself infected with gonorrhœa, or whose vaginal secretion, either in consequence of disordered health or from the presence of menstruation, has become vitiated and so capable of inducing inflammation in the male urethra.

Stages.

The onset of the disease is characterized by a sensation of malaise in the penis, of inconvenience in passing the water, and of pain and irritation along the meatus urinarius, soon followed by swelling and inflammation of the lips of the orifice, tenderness on pressure of the end of the canal, and by the appearance of the symptomatic discharge.

The prominent symptom of the second stage is pain, which may be either burning or shooting, and is felt most severely at the periods of micturition, erection, and ejaculation. The discharge, at first transparent, mucous, and stringy, soon becomes opaque, thick, copious, and fœtid, varying in colour from white to yellow or yellowish-green.

In the third stage there is diminution of the pain and scalding; the discharge gets paler, thinner, and less abundant, and finally ceases altogether. The first stage usually lasts from eight to ten days; the second from eight days to three weeks; the duration of the third is uncertain, as it depends so much on the treatment adopted.

Treatment.

To treat gonorrhœa successfully, the organs must have perfect rest not merely from intercourse but from irritation of every kind. Antiphlogistic measures such as cooling drinks, low diet, tepid baths, cold local applications, and mild aperients, should constitute the treatment during the acute stage. The patient must avoid stimulating drinks, spiced dishes, &c. He should take none but the gentlest exercise on foot, and avoid horse-exercise altogether. A suspensory bandage should be worn by day. When pain has ceased, Copaiba and Cubebs may be given with advantage. These medicines, as M. Ricord has clearly demonstrated, act by combining with the urine and so washing the mucous membrane of the urethra. At the Hôpital du Midi in Paris, the following injections are very commonly and successfully employed, as soon as the acute stage has subsided:

℞ Sulphate de Zinc . . . 1 gramme
   Sous-Acetate de Plomb . 1 gramme
   Eau Destill. . . . 125 grammes.
         or,
Tannin . . . . 50 centigrammes
Eau Destill. . . . 30 grammes

One or other to be injected twice a day, immediately after the patient has passed his urine.

In cases of painful chordee, M. Cullerier prescribes the following pills with great success:—

℞ Camphre . . . 2 grammes
   Extrait de Thebaique . 1 gramme
   Mucilag. Acaciæ . . q. s.

Divide into xx pills: one or two to be taken at bed time.

Among the sequelæ of gonorrhœa may be enumerated:—   *Sequelæ.*

(1.) Accidents produced by violence of inflammation, such as urethritis, balanitis, para-phimosis, inflammation of the lymphatic vessels and of one or more glands in both groins, &c.

(2.) Extension of the inflammation to the testicles, prostate, or bladder.

(3.) Inflammation from transportation of the urethral discharge to some healthy surface, as for instance gonorrhœal conjunctivitis, &c.

(4.) Affections which remain or re-appear after the cessation of the purulent discharge, such as neuralgia of the neck of the bladder or of the urethra, gleet, gonorrhœal rheumatism, &c.

Among the secondary consequences of gonorrhœa, none cause *Gleet;* more trouble to the surgeon and inconvenience to the patient than the affection popularly termed 'gleet,' a name which

56 SKIN DISEASES

includes all indolent discharges issuing from the urethra, and obstinately persisting after the complete disappearance of inflammatory symptoms. In subjects of Lymphatic or Scrofulous temperament, especially where the presence of ascarides in the rectum is a perpetual source of irritation, or again in those who give themselves up to the practice of vicious habits, gleet may exist without any previous acute stage. Want of cleanliness; venereal excess; abuse of alcoholic drinks; over-fatigue in walking and especially in riding; an ulcerated, granular, or constricted condition of the urethra; chronic inflammation of the prostate, vesiculæ seminales, or neck of the bladder; any of these causes may excite a gleety discharge, but in the great majority of cases it orginates in a gonorrhœa which has been either neglected or badly treated.

Its Causes,

Varieties. There are two varieties of gleet; one, belonging to a peculiar constitutional diathesis and to be regarded as a perversion of the secretion of the urethral mucous membrane; a second, which either succeeds an acute gonorrhœa or accompanies some lesion of the urethra. In both varieties the symptoms are almost the same, viz., a mucous or muco-purulent intermittent discharge, staining the linen, usually painless, especially plentiful in the morning and aggravated by the least excess of exercise, diet, or sexual intercourse. When it succeeds an acute gonorrhœa, the discharge is usually white, mucous, semi-transparent and stringy, but becomes under the slightest local or general provocation yellow and muco-purulent. When it is the symptom of some lesion of the urethra, it is ordinarily more or less stained with blood. To foretell the duration or result of a gleet is impossible. It may last for months, whether submitted to skilful treatment or allowed to run its own course without interference. It certainly is far more difficult to manage than an acute gonorrhœa, but, so long as it is not of a purulent character, it is innocuous and non-contagious. As a local accident it is not of much consequence, but unfortunately its subjects are very sensitive, and worry and bother themselves about their discharge to such an extent that they nearly frighten themselves out of their seven senses and become easy victims to the shoals of advertising charlatans and sympathizing friends, who all do their utmost to maintain the unlucky sufferer in his delusion, so long as he is weak and credulous enough to pay handsomely for their christian philanthropy.

A judicious employment of bougies, either plain or coated with some medicinal preparation, will often succeed in curing this troublesome complaint. M. Chassaignac has invented an instrument for conveying the medicinal agent to the part affected, and has used it with great success. Cod Liver Oil, Steel, Bitters, &c. with the use of cold fresh or salt water baths, will sometimes act like magic in effecting a cure when other measures have signally failed. Moderate sexual intercourse is to be recommended rather than proscribed, the frequent erections incidental to a life of enforced chastity being far more hurtful to the complaint than the occasional irritation of sexual indulgence. The patient however must absolutely forbear to gratify his sexual impulses, until the discharge has lost all its purulent (i. e. contagious) properties. In some obstinate cases of gleet, injections of Trisnitrate of Bismuth prove of great service. M. Cullerier is in the habit of employing the following formula with great success:— *Treatment, in the male.*

℞ Bismuthi Trisnitr. . . 20 à 30 gram.
Eau Destill. . . . 200 gram.

Since the introduction of the speculum, affections of the female genital organs have been far better understood and *In the female.*

more successfully treated. Previous to this, the medical man contented himself with merely separating the labia and taking a peep at the vaginal orifice. M. Recamier, who exhumed this much abused instrument from the tomb in which ignorance and prejudice had buried it, thereby conferred an immense boon upon the medical profession; and M. Ricord has been enabled by its aid to establish the diagnosis of female venereal affections with a precision before unattainable. Blennorrhœa in the female includes not only gonorrhœa but also inflammation of the urethra, vulva, vagina, and uterine canal. Without entering into a description of each of these affections, I will merely observe that, making due allowance for the different formations of the genito-urinary organs—the urethra being the seat of mischief in the male and the vagina in the female—gonorrhœal affections of these organs in the two sexes are much the same in point of symptoms, progress, and treatment required. The results of the disease are also in great measure identical; leucorrhœas, vaginal and uterine affections being just as intractable as gleets and other disorders of the urethra, prostate &c. in the male. The same principles of treatment are applicable to both sexes. It is however desirable to bear in mind that injections may be employed of greater strength and with far less fear of evil results in the female than in the male.

B. CHANCRE.

Two kinds of Chancre. There are two varieties of chancre, 1. The soft, non-infecting, or simple; a purely local affection and not followed by constitutional symptoms: 2. The hard, infecting, or Hunterian chancre, followed by a long train of secondary accidents.

That eminent Parisian authority and pupil of Ricord's,

OF CONSTITUTIONAL ORIGIN.

M. Clerc, at whose private institution for the treatment of venereal affections I had the privilege of a lengthened attendance and from whom I received the most kind and friendly attention, recognises the following points of difference between the infecting and non-infecting chancre.

Infecting Chancre	Non-Infecting Chancre	
Has two periods of incubation; one of about 21 days, previous to the appearance of the chancre; a second of about 52 days, previous to the appearance of the secondary symptoms.	Has no period of incubation.	Points of difference between them.
Is indurated, solitary, and accompanied by indolent non-suppurative enlargement of the glands in both groins.	Is non-indurated, multiple, and often attended with suppurating buboes.	
Does not inoculate the neighbouring parts, and is seldom and only under certain conditions inoculable on the same subject.	Inoculates the neighbouring parts, and is readily inoculable on the same subject.	
Has a peculiar form and aspect.	Has a peculiar form and aspect.	
The pus of an infecting chancre, inoculated on a virgin subject, will produce an infecting chancre with subsequent constitutional symptoms.	The pus of a non-infecting chancre, inoculated on a virgin subject, will produce a non-infecting chancre, and is not followed by constitutional symptoms.	
But if the subject has or has previously had constitutional Syphilis, the result (if any) will be a non-infecting chancre, not followed by secondary symptoms.	Even if the subject has or has previously had constitutional Syphilis, the result will be a non-infecting chancre.	
If a man affected with constitutional Syphilis has sexual intercourse with a female having a non-infecting chancre in a state of inoculation, the man will get a non-infecting chancre.	If a woman possessing a non-infecting chancre has intercourse with a man affected with constitutional Syphilis in a state of inoculation, the woman will get an infecting chancre followed by the usual secondary accidents.	

The Non-Infecting or Soft Chancre.

Characters of soft chancre. A soft chancre is an ulcer with a soft non-elastic base, secreting pus which can readily be inoculated and which ordinarily reproduces a soft chancre. Its seat is very variable; but in the male the penis, glans, or prepuce, in the female the fourchette, labia majora, or clitoris, are as a rule its favourite localities. All parts of the skin and many of the mucous membranes are exposed to this disease, but one region, viz. the cephalic, seems to enjoy a marked exemption. The soft chancre has three distinct stages or periods of evolution, viz. the period of incubation, of progress, and of repair. The first is very short and uncertain, so much so that it is impossible to allot any time as its limit. The second, when not retarded by unlucky complications, usually lasts from three to four weeks; and the third is often of merely a few days' duration. When the chancre has run its course, the period of repair commences; the infiltration of the base disappears; the floor of the ulcer becomes clean and covered with fleshy granulations; the borders lose their prominence; and cicatrization commences from the circumference towards the centre or *vice versâ*. The poison of the soft chancre frequently extends to the glands in the groin, and there sets up suppurative inflammation.

Treatment by Caustics, M. Cullerier strongly advocates cauterization as a most efficient means of checking the extension of the chancre, and for this purpose recommends the paste of Charcoal and Sulphuric Acid so much employed by M. Ricord. M. Rollet of Lyons states the advantage of cauterization to be twofold, 1. destruction of the chancre, 2. prevention of the bubo. The following is his mode of proceeding. After wiping the sore clean and dry, he covers it with a layer of some caustic paste of the same size and shape as the chancre, retaining it *in situ* for a couple of hours by means of strips of adhesive plaster. The eschar falls off on the third day, and in about eight or ten days the sore has healed. In small chancres he leaves the paste on for only half an hour; in buboes, from three to four hours. But in the majority of cases of soft chancre there is never, according to my own experience, any need to have recourse to so painful a measure as that of destruction by caustics. Perfect rest of the part, together with cleanliness and soothing or slightly astringent preparations, will prove all that is needful. and by milder means. Ointments should never be employed. When the suppuration is abundant, weak Tincture of Iodine, Decoction of Cinchona or Oak Bark, Vin Aromatique, or some such preparation, should be applied several times a day on a piece of lint. M. Cullerier occasionally employs dry applications, such as powdered Calomel or Alum, and with very satisfactory results. When the chancre occurs in the folds of the anal region or in the lower part of the rectum, emollient applications, such as plugs of oiled lint, should be employed, and the rectum kept clean and empty by the daily use of an injection of tepid water. In ordinary cases the soft chancre is a very mild affection, merely requiring local treatment. General therapeutic measures are not called for, unless it be complicated with inflammation of a gangrenous or phagedenic character, in which case the patient will need both stimulant and tonic medicines, such as Ammonia, Æther, Bark, Quinine, Iron, &c. together with a generous diet. The internal use of Mercury in any case of soft chancre is as unnecessary as it is prejudicial; indeed the old-fashioned indiscriminate way of exhibiting this mineral in all ulcerations of the genital organs cannot be too strongly condemned.

The Infecting or Hard Chancre.

Characters of hard chancre.
Hard chancre stands in the same relation to constitutional Syphilis as the bite of a mad-dog does to Hydrophobia; it is the initiatory symptom, the invariable starting-point of the disease. Commencing either as a pimple or an excoriation, it soon becomes an ulcer resting on an indurated base and involving the whole thickness of the skin or mucous membrane. It is circular, excavated, with clean-cut borders, as though punched out from the surrounding skin. The floor is grey *nacre* or dark red, finely granulated, and devoid either of pseudo-membranous exudation or appreciable suppuration, the only secretion from its surface being a small quantity of clear serous fluid. The base presents a peculiar elastic induration, owing to fibrinous infiltration of its structures. The secondary affection of the glands is constant, but of an indolent nature; suppuration very rarely occurs; and, when it does occur, a simple abscess is formed not inoculable on the same subject.

Such are the characters of a typical case of hard or Hunterian chancre. At the same time it must be remembered that in different cases these characters undergo considerable variations. Into these several varieties of the sore I shall not enter; I will merely observe that, whatever peculiarities it may present, it will always retain the characteristic features of ulceration and induration. The latter commences, according to M. Ricord, towards the end of the first week after contagious intercourse, and attains its maximum about the end of the second or beginning of the third week. After this the surface begins to clean, assumes a healthier colour, and becomes covered with pink granulations. The borders slowly become depressed and cicatrization commences. The induration also subsides, but does not entirely disappear even for some time after complete cicatrization of the sore.

When a patient presents himself with unmistakable hard chancre, what is to be done? Should Mercury be given or should it not? The majority of English Surgeons will most assuredly hold that Mercury in some form or other is indispensable. On the other hand M. Cullerier, whose great experience and success in the treatment of Syphilis entitle his opinion to considerable weight, maintains that Mercury should be wholly withheld. He believes that at this period of the affection it merely retards without warding off secondary symptoms, and that it should be reserved for the cutaneous manifestations of the disease. From what I have seen of his practice at the Hôpital du Midi I am satisfied of the correctness of his views. MM. Ricord and Sigmund have advocated the destruction of the chancre by caustic, provided the patient present himself within the first four days of its appearance; but the results of this plan, notwithstanding the sanction of all the eminent Syphilographers from Vigo to Hunter, are so very dubious that much reliance cannot be placed on its adoption. Slightly stimulating applications with great attention to cleanliness are usually sufficient. These measures, combined with rest and general hygienic precautions, promote the cure of the sore at least as quickly as any other mode of treatment; and this too without taxing the powers of the constitution, which cannot be too carefully husbanded for the subsequent specific treatment of the cutaneous and other secondary symptoms. Indeed in many cases (especially among the population of this metropolis) it is necessary to take more decided steps and to support the patient's strength during the whole progress of the chancre by a course of tonic medicine, such as Quinine or Bark with

Treatment.

one of the Mineral Acids, and a more or less generous diet. When I was House Surgeon at the Lock Hospital, I invariably pursued this plan of treatment and can add my testimony to its efficacy.

SECONDARY VENEREAL AFFECTIONS.
A. SYPHILITIC VEGETATIONS.

Condylomata.

There are two kinds of Syphilitic Vegetations, namely Condylomata and Excrescences. The former, the most common and on account of their contagious character (Hardy) most important of the two varieties, have been described under various names according to their form, structure, and locality. They consist of round or oval elevations, similar in consistence to mucous membrane, and occasionally encircled with a reddish areola. They commence by slight swelling and redness of the part about to be attacked; its epidermis next becomes raised and presents a surface either red and sanious, or covered with a grey mucopurulent secretion which usually emits a most disagreeable sickly odour. When flat and seated on a mucous membrane, they are hardly noticeable; when on the other hand they are raised above the level of the membrane, and especially when they have projecting or overlapping borders, they cannot easily escape observation. Their surface is either uniformly smooth, or else irregular from being fissured or slightly ulcerated. Females and persons of blonde complexion and lymphatic temperament are chiefly liable to these growths. In the former sex their favourite localities are the mammæ, vulva, labia majora and minora; in the latter the prepuce and scrotum; in both sexes the anus, inner and outer surface of lips, tonsils, pharynx,

Their physical characters,

Favourite localities,

pillars of fauces, and tongue. They also occur, though less frequently, on the inside of the thighs, navel, arm-pits, and nasal orifices—in a word, upon all the mucous membranes in immediate contact with the air, and upon all those parts of the skin, which from their habitual condition of warmth and moisture bear considerable resemblance to mucous membranes. Condylomata could not select more awkward situations than most of those just mentioned; the friction to which they are necessarily subject rendering them always more or less painful, irritable, and prone to inflame. Their mode of origin is two-fold; they may appear either spontaneously on a previously healthy surface of skin or mucous membrane, or as the results of the transformation *in situ* of a chancre in a state of ulceration or even cicatrization. and modes of origin.

The aspect and form of excrescences are subject to endless variety. Their colour also varies from that of the natural skin upon which they grow to that of a bright cherry. They may be sessile or pediculated, and have received different names according as Syphilographers have fancied a resemblance between them and other objects, such as cauliflowers, raspberries, &c. While aware of the occasionally Syphilitic origin of these excrescences, we must not forget that the anus and vulva are often beset with growths exactly similar, and that too without the slightest existing or antecedent trace of Syphilis. Excrescences:

Syphilitic or other vegetations, if broad-based, should be destroyed with the Acid-Nitrate of Mercury, Sulphate of Copper, or Powder of Savine. If pediculated, they are best removed with the scissors or écraseur, their site being subsequently touched with Nitrate of Silver or Perchloride of Iron Solution. Their treatment.

B. SYPHILITIC SKIN DISEASES.

Syphilitic skin-diseases.

'Syphilides' is the title given in France to those skin-diseases which form part of the secondary phenomena of Syphilis. Biett was the first to investigate their real nature and to classify them according to their various elementary lesions. The labours of Biett's disciples, M.M. Cazenave, Martin, Legendre, Bassereau, Devergie, and Hardy, together with those of other modern Syphilographers, have done much to facilitate both the diagnosis and treatment of these affections. Before proceeding to notice the several varieties of venereal eruptions, I will briefly describe their common features in respect of colour, form, seat, local and general phenomena.

Their colour;

Their colour is very characteristic, so little resembling anything else that it is usually spoken of as the 'Syphilitic tint.' It is unlike the vinous tint of Scrofulous eruptions or the bright red of ordinary inflammation. French writers often term it 'cuivrée,' an epithet which has received general approval; though at first the *red* tint is usually predominant, and it is not until the full developement of the eruption that the specific *coppery* colour becomes pronounced.

Form;

In form they are nearly always oval, circular, or semi-circular. This character, although not peculiar to Syphilitic eruptions, becomes of some diagnostic value when associated with other symptoms.

Painlessness;

Absence of pain and irritation is the rule, their presence the exception. When a patient with a well-marked Syphilitic eruption complains of either or both of these symptoms, the surgeon may suspect the coexistence of some Dartrous or Arthritic taint.

They are marked by multiformity of character, that is to say, by simultaneous developement on the same subject of many elementary forms of eruption. Nothing is commoner than to find condylomata, exanthemata, papules, vesicles, and pustules coexisting in the same patient. *Multiformity;*

The products of venereal eruptions are peculiar. The scales are smaller, whiter, drier, and more adherent than those of ordinary squamous affections, and are encircled by a whitish border. The crusts also which succeed the scales or the rupture of the pustules are harder, thicker, darker, more fissured and more adherent than those of other cutaneous diseases. The ulcers are generally round, have sharp-cut edges and a greyish floor, and are encircled with a copper-coloured border. The scars left by the ulcers present at first a violet, next a coppery, and lastly a white tint. Their reticulated and flattened appearance together with their peculiar colour facilitates their diagnosis from other scars of non-venereal origin. *Products;*

All parts of the body are liable to Syphilitic eruptions; but, as a general rule, it may be stated that the Papular forms have a predilection for the back of the neck and front of the chest; Condylomata for the neighbourhood of the natural orifices, the mucous membranes, and the most delicate parts of the skin; Psoriasis for the palms of the hands and soles of the feet. *Seat;*

Among the principal concomitant phenomena of these eruptions may be mentioned, ulceration of the throat with alteration of the voice; infiltration of the glands of the neck; neuralgic pains in the head, limbs and joints; iritis; more or less baldness of the scalp, eyebrows, whiskers or moustache; and—at a later stage—various affections of the bones and periosteum. A *and concomitant phenomena.*

certain amount of febrile disturbance usually precedes the eruption and vanishes as the latter becomes fully developed.

Having thus sketched an outline of the general characters of Syphilitic eruptions, I will now enter more into detail and briefly consider their several varieties. As these are exceedingly numerous, it will be necessary first to adopt some system of classification. The simplest appears to be that employed by M. Hardy, who classifies them according to the nature of their anatomical lesions, thus : --

Hardy's classification of Syphilitic eruptions.

1. *Pigmentary.*
2. *Exanthematous.*
3. *Vesicular.*
4. *Pustular.*
5. *Papular.*
6. *Bullous.*
7. *Squamous.*
8. *Tubercular.*

1. Pigmentary.

Pigmentary. This variety of Syphilitic eruption, to which the attention of the medical world was first directed by M. Hardy and subsequently by M. Pillon, is characterized by a well-marked cutaneous mottling of the colour of *café au lait* and very different from the grey tint of Pityriasis. The patches are not raised above the surface of the skin, do not desquamate, and are unattended with heat or irritation. They vary in size from sixpence to a shilling, are usually circular with irregular borders, and being placed side by side cover a large extent of surface. Usually they are distinct from each other; sometimes however they coalesce, especially on the neck, and then present the appearance of veins of marble enclosing spaces of healthy skin, which latter, unless caution be exercised, are liable to be mistaken for the affected parts. The outer borders of these marblings gradually fade so as to blend with the normal colour of the skin. As the disease subsides, the spots become paler and paler, until they finally disappear and the surface resumes its natural tint. The neck is sometimes quite encircled with a chain of these discolorations; at other times the sides only are affected. The disease occasionally declares itself not in mere mottling of certain parts, but in slight pigmentary staining of the whole cutaneous surface. When this occurs, it should be regarded as an important symptom, for it shows how thoroughly the Syphilitic poison has invaded the system. M. Hardy has up to the present met with the disease exclusively among females. M. Pillon has also seen it in certain male patients, whose skins in point of thinness and delicacy resembled those of the other sex. It rarely manifests itself until about the close of the secondary phenomena of Syphilis, and is nearly always a most obstinate intractable affection.

The next variety, the *Exanthematous* (Syphilitic Roseola), is one of the commonest and earliest secondary symptoms. In the subjects of Syphilis this affection almost invariably occurs, but sometimes so slightly, that it may be present and go through its regular course without the knowledge of the patient or surgeon. The former perhaps has his attention directed to it accidentally after a bath, or by the coexistence of some more severe and palpable eruption; sometimes it is first detected by the latter, when the patient comes to consult him for a sore throat or some other inconvenient sequela of Syphilis. It is characterized by irregularly circular patches varying in size from sixpence to a shilling and scarcely, if at all, raised above the level of the skin. These patches vary so much in shape and colour as to suggest the idea of a medley of eruptive elements —the cause no doubt of the numerous appellations bestowed upon the affection by Dermatologists, who have aimed at giving an exact description of the Syphilitic Exanthems. The variations, however, depend merely on the different qualities of the

marginalia: II. Exanthematous. Syphilitic Roseola: Its physical characters,

skins, the parts of the body attacked, and the age of the eruption itself.

The primary tint is slightly carmine, but it may afterwards present a variety of shades from that of a pale rose to that of a brownish copper-colour. When the eruption appears on the belly, it is far more vivid and strongly marked than when it appears on the limbs; as a general rule, the greater the distance from the trunk, the less bright the tint of the spots. The latter are generally very numerous; are unattended with heat, pain, or irritation; and after reaching a certain stage of developement gradually disappear. In recent cases they vanish momentarily under pressure, but at a later period become fixed and permanent.

Progress,
Syphilitic Roseola, though one of the earliest symptoms of the general infection of the system by Syphilitic poison, rarely declares itself before the third week or after the third month subsequent to the appearance of the primary sore. In some rare cases it occupies a period of from two to three months; but in ordinary cases, if judiciously treated, it terminates by resolution in from fifteen to twenty days. Symptoms of general febrile disturbance are apt to precede or accompany the eruption, but of too slight and transient a character materially to affect the patient's health.

and concomitant phenomena.
The possible concomitant phenomena are numerous. It may be accompanied by any of the other Syphilitic eruptions, or by sore throat, loss of hair, neuralgic pains in the head and limbs &c. &c. The coexistence of such complications almost always indicates virulence of infection, and presages the ulterior developement of far more serious mischief than the mere exanthematous eruption. As a disease of the skin, it involves no structural lesion and therefore is of very little consequence; as

OF CONSTITUTIONAL ORIGIN. 71

a symptom of Syphilis, it indicates intensity of the Syphilitic poison and so becomes of far more serious import.

Of the *Vesicular* species there are three varieties, viz. the Eczematous, Varioliform, and Herpetiform. The vesicles, which constitute its characteristic lesion, present almost as great diversity as they do in non-Syphilitic eruptions of the same class. They possess, however, two distinguishing peculiarities, viz. great persistence, and tendency to remain intact for several days.

III. Vesicular. Its varieties.

Syphilitic Eczema consists of little vesicles, sometimes isolated, sometimes grouped together as closely as in ordinary Eczema. These are encircled by peculiar copper-coloured borders, which form by their union large dark-red patches surmounted by the vesicular elements. The liquid contents of these vesicles may either remain intact and be re-absorbed, or burst and so form crusts which leave behind them brown patches of corresponding extent. The vesicles, however, usually retain their integrity longer in the Syphilitic than in other forms of Eczema. Successive relapses of vesicles and crusts may prolong the disease to an indefinite period, but at length they cease and, as far as the eruption is concerned, the patient is cured.

(a) Eczematous.

The second variety (Varioliform), more common than the preceding, presents larger and more scattered vesicles; sometimes they are of the size of a small bean, constituting Bullæ rather than vesicles. Whatever may be their size or shape, they contain a serous fluid, which rapidly becomes purulent, and are encircled with a well-marked copper-coloured and slightly elevated border. At the end of a few days these vesicles burst and are replaced by thick adherent crusts of the true Syphilitic colour and aspect. When these crusts have fallen off, a little eminence remains which after a certain time disappears, leaving

(b) Varioliform.

behind it a copper-coloured stain with a central depression. This in its turn vanishes and the skin regains its healthy condition.

(c) Herpetiform.
The third variety manifests itself under two forms, Herpes Phlyctenoides and Herpes Circinatus. The former is characterized by irregular groups of globular vesicles. The latter presents smaller vesicles placed side by side in regular order and forming either segments of circles or complete circles. Both descriptions of vesicle are surrounded by the peculiar coppercoloured border, the tint of which gradually deepens. At the end of seven or eight days they break and are replaced by very fine scales, on the removal of which the usual specific stains remain, pointing to the Syphilitic origin of the affection. Herpes Circinatus has no favourite locality. It generally makes its appearance within from one to four months after the date of contagion. Like Syphilitic Roseola it is often associated with other Syphilitic manifestations and is attended with similar constitutional symptoms.

IV. Pustular; Its varieties.
(a) Acneiform.
Of the *Pustular* Syphilitic eruptions there are three varieties; the Acneiform, Ecthymatous, and Pustulo-Crustaceous.

The first consists of pustules composed of two distinct parts —a non-suppurating base, and an apex containing a purulent fluid which concretes into a little yellow or brown crust. Each pustule is encircled by a well-marked reddish-brown border. Syphilitic Acne has not, like ordinary Acne, any favourite localities; it may be acute or chronic, but is usually the latter; and it is seldom confluent, even when the pustules are arranged in groups. Two to three weeks may elapse without any change in the pustules; they then break and their contents concrete into little dry, unequal, closely adherent crusts. These when detached reveal either a slight depression or a papular elevation, both always presenting the specific copper-colour. The eruption is apt to be prolonged over several months, especially when the patient has not undergone a proper course of anti-Syphilitic treatment. Indifference as to locality, the depressed and reticulated form of the cicatrix, together with its coppery tint, are the points to be considered in diagnosing between Syphilitic and ordinary Acne.

The second variety, more common and more serious than the first, consists of large usually isolated pustules encircled by a deep red border, but without any induration of base. The pustules quickly concrete into brownish black crusts, and pursue the same course as in ordinary Ecthyma. The favourite localities of this eruption are the scalp and limbs. When the scalp is invaded, more or less baldness is nearly always the result. (b) Ecthymatous.

The third variety, commonly known as Syphilitic Rupia, is by far the most serious of the Syphilitic Pustular eruptions. It consists of groups of very large pustules, forming patches of various sizes and shapes. When these pustules break, their contents concrete into adherent crusts made up of concentric layers which are perpetually increasing in thickness by the effusion and concretion of fresh matter. These crusts are usually hard, greenish-brown or black in colour, rocky, unequal on the surface, and having the appearance of being embedded in the subjacent ulcerations. The latter, when exposed by the detachment of the crusts, are found to be large and deep circular erosions with clean-cut borders and a greyish floor. When the eruption leaves one part to attack another, it is usually termed 'Serpiginous.' (c) Pustulo-crustaceous.

The Pustulo-Crustaceous usually occurs at a much later period than the Acneiform and Ecthymatous varieties, and must be regarded as a tertiary rather than a secondary symptom of

Syphilis. Years, sometimes as many as 15 or 20, may elapse before it makes its appearance. Its presence is always a serious symptom, whether locally on account of the ravages it produces, or generally as denoting a profound and inveterate case of infection. In most cases it is tediously obstinate and unamenable to treatment.

v. Papular.

The next class of Syphilitic eruptions is the *Papular*. This variety is one of the commonest and earliest of the secondary phenomena and often coexists with Syphilitic Roseola. It is characterized by little round flat bodies beneath the skin, forming shiny elevations of the size and shape of lentils. Their colour varies with the stage of the eruption. The tops of the papules are covered with slight delicate scales, and their bases are encircled with a shining white areola. Each papule may undergo successive periods of desquamation. The eruption pursues sometimes a rapid, but more often a slow and insidious course. Its favourite seats are the nape of the neck, face and limbs—occasionally the trunk, especially the back and loins. The small size of the papules and the absence of any ulceration on their surface sufficiently distinguish them from the so-called Syphilitic 'tubercles.'

vi. Bullous.

The *Bullous* form of Syphilitic eruption occurs almost exclusively among newly born infants, whence it has received the name of 'Pemphigus Neonatorum.' Sometimes it manifests itself from the moment of birth, at other times not until a day or two afterwards. It is characterized by the presence, usually on the hands and feet, of violet stains supporting Bullæ of various sizes which contain a yellowish fluid. After a few days these Bullæ burst and are succeeded by superficial ulcerations which soon become covered with crusts. The eruption attacks the robust and healthy no less than the weak and sickly. The infant may seem for a time in good health, but in a few days after the developement of the eruption obstinate diarrhœa and vomiting generally set in and death ensues from exhaustion.

The *Squamous* is one of the commonest forms of Syphilitic eruption and affects the trunk, limbs, palms of the hands and soles of the feet. It presents adherent scales resting on a brownish copper-coloured surface which extends slightly beyond the margin of the scale. There are three varieties of Squamous eruption, viz. Lepra, Psoriasis, and Callosity.

VII. Squamous.

Its varieties.

Lepra consists of slight reddish-brown circular elevations surmounted by fine small white non-imbricated scales. After a time the elevations subside, the scales disappear, and a stain remains which gradually assumes the specific coppery hue. This discoloration fades in its turn and at last completely vanishes. The favourite seats of Lepra are the neck and limbs. It is usually (in Syphilis) a secondary symptom and accompanied with some of the other phenomena belonging to that period of the disease.

(a) Lepra.

Psoriasis, a twin affection with the preceding, is characterized by reddish-brown or copper-coloured elevations covered with non-imbricated scales, much smaller and thinner than those of ordinary Psoriasis. These patches are arranged in segments of circles or ovals; are not usually spread over a large surface; and are separated from each other by intervals of healthy skin. The elevations after a while become depressed and the scales fall off, leaving only slight and transient stains. Syphilitic Psoriasis is frequently associated with other secondary cutaneous affections, especially those of the Papular sort. Its predilection for the palms of the hands and soles of the feet, together with the peculiar colour, limited extent, and thinness of the scales, is generally sufficient to establish its specific origin.

(b) Psoriasis.

SKIN DISEASES

(c) Callosity.

Syphilitic Callosity is also prone to attack the palms of the hands and soles of the feet. It consists of little circular slightly-raised patches surrounded with a well-marked copper border. The epidermis is thickened and indurated, presenting the familiar appearance and pathological condition of an ordinary Corn. It is a very common secondary accident and occurs very shortly after the disappearance of the primary sore.

VIII. Tubercular.

The *Tubercular*, one of the least common and most serious Syphilitic skin-diseases, consists of raised circular tumours of moderately firm consistence, varying in size from a pea to a nut, devoid of pain or irritation, and encircled with the characteristic copper-colour. It belongs strictly to the tertiary class of phenomena, since it rarely appears till after several other forms of Syphilitic manifestations; indeed its commonest victims are patients who have enjoyed an immunity from further symptoms of Syphilis for 15 or 20 years. It attacks all parts of the body, but more especially the face, back, and dorsal aspects of the limbs, and has its seat either in the deep layers of the dermis or in the subcutaneous cellular tissue. Dermatologists have described four varieties according to the nature of the eruption and its mode of termination; viz., the Aggregated, Disseminated, Perforating, and Serpiginous.

Its varieties.

(a) Aggregated.

In the Aggregated form the tubercles are arranged in groups. Some are small, hard, and with apices very slightly raised above the surface of the skin but with bases which seem to contain all the thickness of the tubercle; others (and these are the kind most commonly met with) much larger, varying in size from a pea to a small nut. Their disposition, though sometimes irregular, is usually circular. The tubercular components of the circle may be either perfectly distinct, or else touch each other at their borders and so form an unbroken ridge. The tubercles at first are of a bright red colour, but at a later stage assume the copper hue peculiar to Syphilitic eruptions. Their surface is tense, smooth, and shining. The usual mode of termination is by resolution; the tumours diminish, become scaly on the surface, and finally leave an indelible depressed cicatrix, clearly proving that the skin without being ulcerated has undergone structural lesion. In some cases the centre of the tubercle softens and presents a round ulceration covered by a greenish-black crust. Cicatrization occurs as before, only after a longer interval. The favourite localities of this eruption are the face, lips, chin, and alæ of nose.

The Disseminated form is characterized by the development of round dark-red shining tubercles, varying in size and dispersed possibly over several parts of the body, but having a predilection for the face, trunk, and upper extremities. At a later period they are covered with delicate white scales; they do not ulcerate, and leave behind them only transient marks. The eruption is unattended with pain and irritation; is a secondary lesion, occurring usually several months after the primary sore; and is frequently associated with other eruptions, especially the Papular. It affects a circular mode of arrangement. (b) Disseminated.

The Perforating variety, far more important and serious than either of the preceding, usually commences by the appearance upon different parts of the body, but more especially the face and limbs, of tubercles of various sizes, numbers, and shades of colour. These tubercles, sometimes isolated, sometimes in groups of three or four, are dispersed irregularly and never cover the whole surface of the body at once, although they may do so by several successive eruptions. At first full and hard, they afterwards soften; the skin becomes thinned and at last (c) Perforating.

perforated, presenting a more or less deep ulceration with clean cut edges and a greyish sanious floor. Black jagged uneven crusts, resembling those met with in Pustular Syphilitic eruptions, wholly or partially cover these ulcerations and sometimes attain considerable size and thickness. These crusts from time to time become detached and again renewed; the subjacent ulcerations meanwhile increase, not in width but in depth, and destroy all the tissues in their way. The cartilages, nay even the bones, present no obstacles to their ravages. When the face is attacked, the horrible disfigurement renders the unfortunate sufferer an object of disgust to himself and of aversion to his fellow-creatures; the constitution seriously suffers from the intensity of the Syphilitic cachexia; and, if the patient recovers, it is at the cost of an ugly deep permanent scar.

Perforating tubercle rarely makes its appearance till about eighteen months or two years after the primary sore. The prognosis is unfavourable; the disease being tedious and apt to spread, and the constitutional cachexia rendering it very unamenable to treatment.

(d.) Serpiginous. Serpiginous tubercle, like the preceding variety, is characterized by a tendency to spread; but differs in its mode of progression, increasing in width rather than in depth. It commences by several tubercles arranged in various forms and often combined with pustules. The tubercles soften, the pustules break, and both are succeeded by superficial ulcerations covered with thick black jagged crusts. After a certain time these crusts get loose, fall off, and reveal uneven violet scars. Meanwhile other tubercles or pustules appear on neighbouring parts and ulcerate; then perhaps a third crop spring up in some more distant situation and go through the same stages. The eruption is surrounded by a well-marked copper border, and nearly always affects an oval or circular mode of arrangement. The special character of this particular variety is its insidiously progressive mode of extension, whence its name Serpiginous, from Latin *Serpo*, to creep. Although a tertiary affection, it is less serious than the Perforating tubercle, yields more readily to treatment, and is not attended with such indelible disfiguring scars.

Treatment of Syphilitic Eruptions.

Four men, once upon a time, fell into bad company and contracted Syphilis. No. 1 had a Chancre; No. 2 a Chancriform Erosion;* No. 3 a blushing Roseola; No 4 a visit from an old attached Psoriasis. Like sensible fellows they put themselves into their surgeons' hands who, strange to say, although brought up in medical schools and theories quite as various as their patients' symptoms, were unanimous on this occasion at least in prescribing Mercury to their four invalids. Their victims demurred to this mode of treatment. *Indiscriminate administration of Mercury a practice well-nigh obsolete,*

'I say, Doctor, if I consent to take this horrid Mercury, will you promise me a thorough cure?'

'My dear fellow, I'll do my best; but neither Mercury nor I are infallible.'

'And, if these blessed scales come back again, what on earth am I to do?'

'Come back again!' Echo replies, 'Scales or no scales, I'm always delighted to see you.'

'Many thanks! and what will you give me then?'

* By 'Chancriform Erosion' Mr. Diday means a *modified* Chancre resulting from the contagion of a *secondary* accident or lesion—in contradistinction to the *genuine* Chancre produced by contact with a *primary* sore. On this point see pp. 89-90.

'Don't anticipate the evil day. Go out of town; take great care of yourself like a good boy; and give me a look in whenever you're passing.'

M. Diday has given this little illustration of the old-fashioned mode of treating Syphilis, a fashion which it is to be hoped is now well-nigh extinct. Mercury certainly does cure many Syphilitic lesions, especially the more serious and destructive; yet many of the earlier and less grave phenomena are amenable to simple, non-specific treatment. A conscientious medical man will always bear in mind the three following facts:—

<small>and open to grave objections.</small>

I. That serious inconveniences attend the exhibition of Mercury, among which may be mentioned diarrhœa, ulceration of the gums and consequent loss of teeth, impaired nutrition, emaciation, asthenia, tremors, and general symptoms of nervous depression.

II. That it does not prevent the developement, impede the return, or shorten the duration, of Syphilitic lesions. This has been proved by Mr. Gore of Limerick (see *Lancet*, Sept. 11, 1858); Mr. de Meric in his Lettsomian Lectures on Syphilis, 1858; M. Diday of Lyons; Signor Palasciano of Naples; M. Buzinct, and others.

III. That, even if through excess of caution or uncertainty of diagnosis the exhibition of Mercury has been deferred, its adjournment will entail no serious ill effect upon the patient. Numerous experiments made by the best Continental authorities have settled this point beyond all doubt.

<small>Non-Mercurial treatment the safest.</small> Non-Mercurial treatment, i.e. by simple cleanliness, attention to hygiene and the exhibition of tonics, if needful, may not be so quick but is far more safe than the old system. Ricord very justly condemns the exhibition of Mercury in cases of

doubtful Chancre. 'Don't interfere,' is his advice to surgeons, 'but leave nature alone and wait the spontaneous developement of the malady. If the diathesis exist, before many weeks are over, its presence will be clearly demonstrated. Your course will then be plain. And this, let me tell you, is no mere trifle; for that a man should know whether he has or has not had Syphilis is quite as important as that you should be able to decide on the propriety or impropriety of giving Mercury.' A patient may become tired of bachelor life and wish to marry and settle down. He has been under treatment for the old, old story and has, as usual, been repeatedly salivated. Should he, or should he not, marry? The surgeon is placed in a very delicate and responsible position, and in giving an opinion must be guided by a careful consideration of his patient's present state and past history. If the disease has *not* been subjected to specific treatment—in other words, if it has been left to run its own course, and no fresh symptoms have for the last eight or ten months appeared—no just cause or impediment exists why the two persons, &c. &c. The disease has worn itself out and is now a thing of the past to be associated with Richmond dinners, pets of the Ballet, and other quondam diversions of bachelor life. If the patient has past forty the Rubicon of middle life, is intractable and impatient of rules and restrictions, Mercury is indispensable; as sooner or later, whether the manifestations be slight or grave, it must be exhibited if he is to derive any benefit at all from medical treatment. Should headache, muscular lassitude, mental prostration, or gastro-intestinal disturbance supervene, at once suspend the Mercurial treament and exhibit Iodide of Potassium or Chlorate of Potash combined with preparations of Iron or Quinine. but in certain cases insufficient.

The first Syphilitic eruption is *par excellence* the symptom Therapeutic which imparts an idea of the future sequelæ, and is usually of the same nature all over the body. If the patient has taken Mercury either before or during the existence of the Chancre, it is wise to continue the same treatment. If he has had the good fortune to have escaped Mercurial discipline, the course is plain. Has he a Roseola? Wait. Has he a squamous, vesicular, or pustular eruption? Give Mercury. Has he a papular eruption? Wait, but watch.

indications to be derived from the nature of the first,

The second eruption demonstrates the propriety or impropriety of the previous treatment, according to the nature of its lesions and the interval which has elapsed since the first eruption—considerations which will likewise aid the surgeon in deciding whether to give or withhold Mercury on the present occasion. If the second eruption is the same or of a milder character than the first, specific treatment will not, as a general rule, be requisite; but should the intervening period have been short; should the eruption be of the squamous, vesicular, or pustular kind; should dysphonia, iritis, or any symptoms of hereditary infantile Syphilis be present; then the necessity for Mercurial treatment is clearly indicated.

and of the second eruption.

How long should Mercurial treatment be continued? M. Diday tells us that his long experience tends to confirm the wisdom of his old master's plan. Dupuytren used to lay down the following rule:—Pursue treatment for the same space of time *after*, that it has taken to *obtain*, the disappearance of the lesion. As a precautionary measure, the exhibition of Mercury is useless. Many patients who have had Syphilis come and tell us that they are thinking of getting married, and wish to know whether a course of Mercury would not under these circumstances be advisable. They have no symptoms, no sore throat, no spots. To this delicate query M. Diday replies: 'Mercury is not an antidote; it cannot cure invisible lesion

How long Mercury should be given.

OF CONSTITUTIONAL ORIGIN. 83

If the virus is still puissant but dormant, *requiescat in pace.* When in its own time the disease bursts forth, then we can meet it face to face and combat it with appropriate weapons.'

There are three ways of administering Mercury, viz. by the mouth, by inunction, and by the vapour bath. Modes of administering it.

The first of these plans is attended with so many disadvantages (such as affection of the mucous membrane of the mouth, intestinal canal, &c.) that it has long been falling into comparative disuse. Mercury given in this way, Mr. Henry Lee justly remarks, has first to be absorbed into the blood and carried the round of the circulation, perhaps more than once, before it comes in contact with the affected structures. By the mouth.

By introducing it into the system through the skin by means of friction, its deleterious action upon the internal organs is in great measure avoided. The amount absorbed penetrates all parts of the body quite as thoroughly, without being conveyed directly to the liver, as it is when swallowed into the stomach. But, although the external inunction as compared with the internal exhibition of Mercury has very decided advantages, it is still not without a drawback in the trouble and personal discomfort which it entails. By Inunction.

Fumigation of the surface of the body by the volatilization of certain Mercurial preparations has all the advantages with but few of the disadvantages of the other two methods. While I held the office of House Surgeon at the Lock Hospital, I had abundant opportunities of testing the merits of the several methods; and I am convinced that in cases of constitutional Syphilis, where it is desirable to affect the system with Mercury, this end cannot be attained by any means at once so effectual convenient and safe as the Calomel vapour bath, in the simple form devised by Mr. H. Lee and manufactured by Messrs. By Fumigation.

Superiority of this last method.

G 2

Whicker and Blaise of St. James' Street. It is clean, simple, and easy of application; its strength can be regulated to the greatest nicety; its action can be maintained, with little or no inconvenience, for almost any length of time; and it makes far less demand upon the patient's vital powers than either of the other plans. Moreover, fumigation brings the medicinal agent into direct contact with the cutaneous disease.

Treatment of tertiary symptoms. The so-called tertiary symptoms, comprising tubercular, pustular and ulcerous affections of long standing, require a combination of Mercury and Iodine. The best mode of obtaining the joint effect of these remedies is to give them in separate doses and at different times—the Iodine in the form of Iodide of Potassium every morning in a dose of from ten to fifteen grains in some bitter infusion; the Mercury in the form of pill every night. In tubercular affections, more especially in the perforating variety, M. Hardy recommends the following combination:—

℞ Eau Destill. . . . . 250 gram.
   Iodure de Potass. . . . 16 gram.
   Bi-iodure de Mercure . . ·1 à ·5 centigr.
One tablespoonful to be taken daily in a cup of Tisane.

Emollient lotions and poultices are often necessary for detaching the crusts. When however these are very hard and cover deep ulcerations, the surgeon's attention must be directed, not to the detachment of the crusts but to improving the general health. The crust forms the natural protection of the ulcerated surface, and its removal in such cases simply retards the healing of the latter.

In some cases of Perforating and Serpiginous Tubercle, cauterization with Nitrate of Silver or the application of the following ointment will be found to accelerate the process of cicatrization :—

℞ Axonge . . . . . 30 gram.
   Proto-iodure de Mercure . 5 centigr.

M. Diday tells us that long experience has satisfied him that tertiary Syphilis prevails far more extensively amongst the upper than the lower classes. Patients, who lead sedentary luxurious lives and spend their nights in heated gas-fouled and tobacco-scented rooms, are under the most favourable circumstances for acquiring the Syphilitic diathesis in its most intractable form. The same Syphilographer asserts that tertiary Syphilis must not be regarded as one of the periods of evolution of that disease but as a distinct pathological condition, occurring only when the specific poison has gained a permanent instead of a transient hold upon the system. Their clinical characters, and the different treatment they require, together with their non-contagiousness, are radical points of difference between the tertiary and earlier manifestations of Syphilis. In the former we must rely on careful hygienic rather than on medical treatment. We must treat the disease not directly, but indirectly—by improving the general health. Change of air, food, abode, and occupation, will usually do far more for these cases than the most prolonged and most judicious course of medicine.

Diday's views of tertiary Syphilis.

General Hygiene.

Hygiene plays a most important part in the treatment of the cutaneous and all other secondary manifestations of Syphilis. Our great aim should be to put our patient in the most favourable circumstances possible for maintaining or improving his general health. If the latter is tolerably good, it will be no slight therapeutic achievement to keep it so. If (as is nearly always the case) his constitution is more or less enfeebled, we must do the utmost which our experience and skill can suggest to renew his impaired vitality, and so enable

Hygiene.

him both to eliminate the poison and resist its action. The principal hygienic precautions, which the subject of syphilis should observe, are the following:—

Food. 1. Sufficiency but plainness of food. The diet should consist of ordinary butcher's meat and cooked vegetables, varied with fish and fowl; no game, made-dishes, preserved meats, hot pickles, sweets or pastry. But of plain wholesome nutritious food they should be encouraged to eat heartily, or they will lack strength to bear up against their disease.

Air. 2. Plenty of fresh air. Patients with a vitiated state of blood should live in large, light, sunny, well-ventilated rooms. If they live in the country, all the better. If they live in a large town and, still worse, in a confined street or square, they must get out of it whenever they can. It is one of the blessings of these days of cheap Railway-Excursions, that a trip into the country is a luxury now within the reach of Lazarus as well as Dives. The majority of patients need to be reminded that fresh air and cubic space are not less indispensable by night than by day.

Sleep. 3. 'Early to bed and early to rise' is a maxim which should never be forgotten. Thoughtless youths, the pet subjects of Syphilis, instead of 'lengthening their days by taking some hours from night,' had much better reverse their habits, and lengthen their nights by taking some hours from day.

Exercise. 4. Exercise. Unless it is imperatively necessary, the patient should never be kept confined to the house. Mental and physical depression will otherwise be superadded to his existing disease. Nothing contributes so much to retard recovery as want of occupation and a mopish sedentary style of life. The patient should be allowed to live as far as possible in his usual manner and among his usual companions. Moderate and steady exercise in the open air will sometimes induce that change in the system which, when aided by well directed treatment, will cure even the most obstinate Syphilitic affections.

5. The old saying that 'a burnt child fears the fire' certainly does not hold true in Syphilis. A patient, who has got cured of his chancre, skin disease, or whatever he may happen to have had, flatters himself that he has benefited by his lesson, and that he will not be such a fool as to expose himself to the chance of infection a second time. But human resolves are weak and sexual passions are strong, and the result is that, in Syphilis at any rate, past experience is a poor safeguard against future mishaps. The subjects of this disease are generally young men, and at their time of life strict chastity involves a self-control greater than most can command, while marriage is for a variety of reasons too often impracticable. *Continence.*

6. Sea-bathing, a visit to some well-conducted Hydropathic establishment, or a course of natural mineral waters, often prove powerful aids to the surgeon in accelerating and confirming the cure of obstinate Syphilitic eruptions. For this purpose the Spas of Aix-la-Chapelle, Aix-les-Bains, Bareges, Bagnères de Luchon, Uriage &c. on the Continent, and Harrowgate in our own country, may be highly recommended. Another advantage of the Sulphurous waters is, that their use generally settles any doubt respecting the precise nature of a cutaneous affection which is not attended and confirmed by the usual phenomena. If it be of Syphilitic origin, the waters will soon evoke some other characteristic symptom and so reveal its true nature. *Sea-Bathing, &c.*

Attention to hygienic principles such as those I have above mentioned, together with habits of most scrupulous personal cleanliness and avoidance of fatigue and excitement, cannot be too strongly enforced on the patient. His neglect of such precautions often prolongs the duration of the disease far beyond its natural limits, and this too in spite of any amount of skill and care bestowed upon him by his medical attendant.

CONTAGIOUSNESS OF SECONDARY SYPHILITIC LESIONS.

In December 1858, M. Langlebert published in the *Moniteur des Hôpitaux* a memoir, in which he promulgated and supported his novel doctrine of the contagiousness of secondary Syphilis. Stated briefly the doctrine is this: —

Doctrine of the contagion of secondary Syphilis

1. That the source of true Syphilis, with its long train of primary and secondary accidents, is by no means limited to the sexual act, i. e. to contact with a *primary* sore; but that the disease may be propagated also by its *secondary* lesions, such as condylomata, fissures, &c., and even (in virulent cases) by the blood and bodily secretions (saliva, sweat, &c.) of an infected person.

2. That true or constitutional Syphilis, even when caused by a secondary lesion, always begins with an indurated chancre—the chancre being situated on that part with which the virus of the secondary lesion came in contact.

now well-established.

This doctrine at first met with great opposition, but in 1859 the Academy of Medicine unanimously condemned the old theory of the non-contagiousness of secondary Syphilis; and, on the memorable occasion of its delivering this judgement, the great Ricord himself renounced the opinion which he had held for 25 years, and acknowledged the truth and importance of Langlebert's discovery. It has since been confirmed by the multiplied observations and experiments of MM. Rollet,

Viennois, Fournier, Diday, and several other French surgeons of distinction, so that M. Langlebert's doctrine must now be accepted as one of the great fundamental truths in the etiology of Syphilis.

The evidence adduced in its support is that of carefully observed cases, in which true Syphilitic infection could be attributed solely to contact with a secondary lesion, no primary sore existing at the time on the infecting person. For instance, a Syphilitic patient can unquestionably trace his chancre to a single occasion of sexual intercourse with a woman, who, as far as her primary sore was concerned, was perfectly cured, but who at the time of connection had some condylomata on the vulva. Or perhaps he has a chancre on his lip, and it can be proved that he caught it from a woman whose only Syphilitic lesion at the time or times of intercourse was a condyloma or fissure on her own lip, or perhaps an excoriation on her tongue. In fact M. Langlebert proves incontestably that the sources of Syphilitic infection are far more numerous than are generally supposed; that the power to infect does not cease upon the cure of the primary sore, but is reproduced (only with progressive decrease of virulence) in the secondary and even the tertiary lesions. *Evidence by which it is supported.*

So much for the sources of infection. Let us now turn our attention to its results. As I said before, in all cases of constitutional Syphilis, whether arising from a primary or a secondary accident, an indurated chancre is the first manifestation of the disease. Now this chancre, though essentially the same, presents certain minor points of difference in the two cases. As compared with ordinary chancre, the one produced by inoculation from a secondary lesion—the Chancriform Erosion, as Diday calls it—has (1.) a longer period of incubation; (2.) it *Difference between results of infection from secondary and from primary lesions in*

<small>respect of Incubation, Physical Characters,</small> appears as a dry superficial indolent copper-coloured papular elevation, with much less induration of base, the surface exuding a very small quantity of serous fluid and being covered by a thin crust or scale which from time to time is shed and renewed. I would here observe, that it is this papular character of the chancriform erosion which renders it so liable to be mistaken for a condyloma. The mistake will generally be avoided by bearing in mind that condylomata are wont to appear in crops simultaneously; are without induration (unless inflamed); and do not, like chancres, involve enlargement of <small>and Virulence.</small> the adjacent glands. (3.) Its virus is less strong and less active, as though it had lost somewhat of its infecting power by its passage through the system, and, as a natural consequence, the secondary symptoms which in time succeed it are less severe and less inveterate.

From the milder nature of the lesion in these chancriform erosions, it is easy to see that they present very little obstacle to sexual intercourse and must prove fertile sources of Syphilitic infection. A man blessed with a Hunterian chancre finds the pain incidental to the sexual act (supposing him villain enough to attempt it) far too great for him to venture to repeat it; but the owner of a chancriform erosion, finding it almost or altogether painless, and probably being unaware of its infecting power, continues his homage at Venus's shrine with little or no diminution of ardour.

With such multiplied facilities for sowing broadcast the seeds of Syphilis, the only wonder is that so rank and deadly a weed should not have thrown out its roots, even wider and deeper than it has done, among all classes of society.

SCROFULOUS ERUPTIONS.

SCROFULA has been recognised from time immemorial as one of the most serious maladies to which the human race is liable. Every nation, every class in society, each sex, and all ages alike furnish it with victims. The skin is far from being the only seat of its visitations; the mucous membranes, glands, bones, and viscera, are none of them exempt from the ravages of this fatal disease. *Scrofula:*

Scrofula is a constitutional affection, non-contagious, usually hereditary, intensely chronic, manifesting its existence by an assemblage of morbid phenomena, which however varied in seat possess the following properties in common—viz. fixity; tendency to invade the tegumentary, lymphatic, and osseous systems, and to cause infiltration and ulceration of the tissues attacked. It is further characterized by peculiarities in the physical conformation of the body. As a rule, the back of the head is unduly developed, while the forehead is low, the neck short, and jaws large. The scrofulous physiognomy presents two different types. In one the complexion is fresh and clear, the eye bright and animated, the features delicately chiselled, the hair fine and luxuriant, the limbs rounded, and the general appearance that of perfect health. In the other the complexion is pale and muddy, the eye lacks lustre and animation, the hair is coarse and scant, the features wanting in delicacy, the body lean and angular. Scrofula also appears to *Its general characters; and influence upon* *Physical conformation,* *Physiognomy,*

Stature,

exercise an unfavourable influence over the stature; subjects of twenty often look about fifteen years of age; others may have their full height, but are badly built and put together. The chest is flat and more or less quadrilateral, the belly prominent, and the limbs out of proportion to the rest of the body. The vertebral column is apt to be curved in various directions. The digestion in one may be quick and easy, in another slow painful and attended with flatulent eructations.

Physical, mental, and moral temperament,

Some are quick, active, and industrious; others slow, idle, and lazy. In some the intellectual faculties are keen and brilliant; in others they are so obtuse and muddled, as to place their possessors on a par with idiots and imbeciles. Some are choleric, hasty, and selfish; others mild, patient, and generous.

and sexual impulse.

In some cases the sexual passions show signs of activity at a very early period; in others they are dormant until beyond the age of puberty. Before proceeding further, it will be useful to give a brief *résumé* of the Scrofulous manifestations, and for this purpose I cannot do better than divide them, as M. Bazin has done, into four distinct periods.

Phenomena of its different stages:

Of the first,

Scrofula generally makes its first appearance between the ages of five and fifteen, in superficial affections of the skin and mucous membranes with sympathetic enlargement of the nearest lymphatic glands. The primary affections usually consist of excrescences or vegetations, termed on the skin *verrucæ* or tubercles, on the mucous membranes *polypi*. Eruptions of the scalp and enlargement of the glands, which inflame and suppurate as the disease progresses, constitute as it were intermediate phenomena between the first and second stage. The superficial cervical glands may be affected on one or both sides; but the disease rarely limits itself to the surface, being wont to extend to the deeper glands and thus give rise to abscesses, sinuses, fistulæ, &c.

The second period presents tegumentary affections of a deeper *Of the second,* and more serious character than the preceding. The most troublesome lesions of this period are those which the disease entails when seated in the mucous membranes. Thus we get Scrofulous ulceration of the bowels, causing obstinate and exhausting diarrhœa; erosions of the neck of the uterus, causing leucorrhœa, &c.

To the third period belong affections of the bones and joints *Of the third,* —periostitis, inflammatory enlargement, caries, and necrosis of the bones; 'white swelling' of the joints, with more or less disorganization of their synovial membranes and cartilages. The features also now begin to change and assume a sickly pale colour; the constitution becomes gradually undermined; albuminuria and diarrhœa are apt to set in; loss of flesh ensues, and finally death, from progressive exhaustion.

The fourth period presents parenchymatous and visceral *Of the fourth.* affections which frequently coexist with lesions of the last period—viz. tubercle in the lungs, brain, and other viscera; fatty, albuminous, and tubercular degeneration of the liver, kidney, spleen, &c.

With the latter stages of this fatal disease I have nothing to do, my concern at present being with Scrofula in its relation to the skin.

Scrofulous skin diseases are exceedingly simple. Viewed *Scrofulous skin diseases:* comprehensively they are nothing more than the ordinary skin diseases which have been described in the early part of this work—squamous, papular, vesicular, &c.; only occurring in Scrofulous subjects they present certain peculiarities—the peculiarities, that is to say, which characterize all structural lesions in this diathesis. In the first place a Scrofulous eruption, *Their distinguishing* as contrasted with the same kind of eruption in a Dartrous or

94   SKIN DISEASES

character-istics, Arthritic subject, will be found not to shift about but to keep to one spot. Of course it may appear in two or more parts of the body at the same time; what I mean to say is that, whether the localities of its choice be one or many, to these localities it manifests a most obstinate adherence. A second point of contrast is the tendency of Scrofulous eruptions to spread from the superficial to the deeper layers of the skin, and to form patches of ulceration which, if not checked, gradually extend both in width and depth and destroy all the tissues involved in their progress. Again, the process of repair in Scrofulous subjects being very defective, these eruptions often leave behind them much disfigurement in the shape of deep jagged adherent scars. They are seldom unaccompanied by more or less secondary affection of the adjacent lymphatic glands, which may become merely enlarged and indurated, or may take on inflammation and suppuration. Lastly, being of a chronic insidious asthenic character, they are attended with but little febrile disturbance, and scarcely any of that local pain, heat or irritation, which form so troublesome a feature in other eruptions. Local measures, moreover, are of little avail, the only way of reaching the local disease being to combat the constitutional evil; and, as this latter is often so virulent and so deeply seated as to defy the best directed efforts, there are necessarily many Scrofulous eruptions on which medical skill can make little impression.

As I observed above, the majority of skin diseases to which scrofulous patients are liable are in all essential points identical with the ordinary Dartrous and Arthritic affections already described, and therefore do not need to be described again. Certain of them however become so much exaggerated in, or are so specially or exclusively incidental to, the Scrofulous diathesis as to merit separate consideration. For the sake of

OF CONSTITUTIONAL ORIGIN.

convenience I shall arrange them under three heads, according as their seat is in the superficial or in the deeper layers of the skin or in the mucous membranes. *and chief varieties.*

A. SEATED IN THE SUPERFICIAL LAYER OF THE SKIN.

Scrofulous people, especially in early life, are liable to unusually obstinate *chilblains* on the hands and feet. Their immediate cause in most cases is great and sudden alternations of temperature, as for instance holding the hands and feet close to a warm fire when they are numbed with cold. The more remote cause is depressed vitality and feeble circulation of blood. Chilblains consist of chronic congestion with swelling, sometimes amounting to inflammation, either of the skin alone or of the skin and subcutaneous tissue. When warm, they are of a dull red colour and generally accompanied with much itching and tingling; when cold, they become of a dark venous or livid hue. If neglected, they are apt to 'break,' as the expression is, i. e. to ulcerate, and give rise to troublesome indolent sores. Besides the general treatment proper for Scrofula, which will be described hereafter, there are certain special precautions to be observed for the prevention and cure of this troublesome and not unfrequently disfiguring affection. The circulation being feeble, the diet should be generous and comprise a due proportion of alcoholic stimulants. Brisk outdoor muscular exercise should be enforced, the extremities at the same time being protected to a far greater extent and with warmer clothing than is usually the case. Children troubled with a tendency to chilblains should not be allowed to run about with arms and legs bare; these are the very parts which in them need to be kept warmest. Where chilblains already exist, friction with a stimulating lotion such as the *Linimentum*

Chilblains: their cause,

and treatment; a. generally.

b. locally.

Ammoniæ or the *Linimentum Camphoræ Compositum* of the British Pharmacopœia is the best local treatment. If they have ulcerated, the sores should be dressed with Zinc or Resin ointment, and the hand or foot affected should be wrapped up warm in cotton-wool and kept in a raised position to facilitate the return of blood towards the heart.

The other erythematous affections of the skin besides chilblains, occurring in Scrofulous subjects, present no peculiarities which call for notice.

Of the papular class, there is one which is apt to attack Scrofulous people so commonly and in so aggravated a form, that it would be an omission to pass it by without some consideration; I mean Acne.

Acne and its varieties.

Acne consists in obstruction, distension, and inflammation of the sebaceous follicles of the skin, and appears first as isolated hard red pimples which may or may not go on to suppuration. This is *simple* Acne. Oftentimes, from the accumulation of dirt, sweat, grease, and epithelial scales, the summit of a pimple presents a little black speck. When this is the case, the eruption is specified as *Punctata*, from *Punctum*, a point. When the patch of skin on which the inflamed follicles are situated assumes a permanent rose-red colour, and, together with the subcutaneous cellular tissue, becomes thickened and indurated, the eruption is called *Rosacea*. The two first varieties attack both sexes from the period of puberty to about 35 years of age, their favourite localities being the face, back, shoulders, and upper part of chest. The latter variety attacks chiefly elderly women, and is almost exclusively confined to the nose and cheeks. The subjects of Acne have almost always pale muddy complexions, and their skins are thick and greasy to the feel. The Acne Rosacea of elderly people may be said to be incurable; but the other two forms incidental to earlier life, though most obstinate eruptions, will always get well in time, though they often take months and even years to do so, varying in severity, now almost disappearing, now again returning worse than ever. Though the general health may be good (as good, that is to say, as is compatible with the Scrofulous diathesis) the disfigurement of an Acne eruption is very great, and the subject of it is pretty sure to invoke medical assistance on behalf of his personal appearance. This leads me to consider the treatment most appropriate to the affection.

Our first endeavour will be in each case to ascertain, if possible, the immediately exciting cause. In some cases this will be found to be intemperance in eating and drinking, especially when conjoined with sedentary habits, a sluggish state of the bowels, and the use of strong coffee, tea, and tobacco. Here the therapeutic indications are plain enough. Exercise, abstemious diet, with avoidance or at any rate restricted use of the articles just specified, and attention to the state of the bowels, will probably effect a cure. In females, especially actresses, ballet-dancers &c., I have frequently traced the disease to one of two causes, viz. either the habitual use of irritating cosmetics or suppression of the menstrual function; and by combating the cause I have generally succeeded in restoring to my patient a clean skin. In persons of both sexes from about eighteen to thirty years of age, there is reason to suspect that a not uncommon though rarely acknowledged cause of Acne is masturbation. Why this vicious habit should give rise to this or any other form of cutaneous disease, I confess myself wholly unable to explain; I can simply say that I have known some cases and heard of others in which the eruption came,

Treatment.
a. Preventive.

continued, and disappeared coincidently with the establishment, persistence, and relinquishment of the habit.

b. Curative. But supposing we are unable to discover any exciting cause of the eruption, what is to be done? As there is no known specific for Acne, all that we can do is to keep the general health in as perfect a condition as possible, and adopt certain local measures which experience has shown to be generally beneficial. If the eruption is acute, the local treatment should be soothing and consist of emollient lotions, such as warm milk, emulsion of bitter almonds, starch and glycerine &c., applied at frequent intervals during the day. If on the other hand it becomes chronic, the papules being indolent and making no progress either towards resolution or suppuration, the local treatment must be stimulating. The skin must be well washed each day with common yellow or soft soap and rain or distilled water. A little spirit of wine may with advantage be added to the water to facilitate the removal of the greasy cutaneous secretion and keep the orifices of the sebaceous glands clear. After the soap has been well rubbed in with the fingers and again washed off in clean water, the skin should be dried by brisk friction with a rough towel. The friction, so far from aggravating the eruption, has just the contrary effect. By stimulating the capillaries of the part to a healthier and more vigorous action, it hastens the eruption to a favourable termination. Local stimulus may be further applied in the way of lotions. Great benefit often ensues from touching the summits of the larger and more indolent papules or pustules with the smallest possible quantity of blistering fluid or of strong Solution of Iodine, such as the following:—

Local applications.

℞ Iodinii
   Potass. Iodid. . āā ʒj ⎫ Misce.
   Sp. Vini Rect. . ʒj ⎭

The *Unguentum Hydrargyri Nitratis* (undiluted) of the British Pharmacopœia, or a Solution of Corrosive Sublimate, is a most useful local application in these cases. The latter is best applied, immediately after washing the face, in the form of lotion with a little Rectified Spirit and Muriate of Ammonia, thus:—

℞ Hydrarg.Bichlor.
Ammon. Hydrochlor. . āā gr. viij
Sp. Vini Rect. . . ʒiv
Aquæ Rosæ . . ad ℥iv
} Misce.

A useful and at the same time pleasant lotion may always be extemporized by mixing equal parts of Eau de Cologne or Lavender water with common water.

In chronic indolent Acne, especially in that form of it termed *Punctata*, the best and generally the only plan is to evacuate the larger and black-headed follicles by mechanical pressure between the finger-nails. The sebaceous matter squeezes out, and the inflamed follicle relieved of its contents soon regains its normal size and condition.

Another of the Papular skin diseases, to which Scrofulous subjects are liable, is *Molluscum*. This is a very rare affection, and characterized by the eruption of little tubercles of various sizes containing matter identical with pulmonary tubercle. It occurs not unfrequently in children, the usual seat being the face or neck; it never leads to any bad results, the worst of the eruption being its disfigurement. The tubercles generally remain stationary, without undergoing any appreciable change. Sometimes, however, they inflame; sometimes soften. In either case their contents become liquefied and escape through the orifice at the summit, either spontaneously or under pressure. They then heal with or without a scar according to the nature of the previous inflammation. The best treatment consists in

Molluscum.

stimulating applications to the skin, with evacuation of the contents of the tubercles by pressure between the finger-nails.

B. SEATED IN THE DEEPER LAYERS OF THE SKIN.

Lupus, and its varieties. Thus far I have considered Scrofulous affections which present comparatively slight lesions and are limited to the superficial layers of the skin. I now proceed to notice certain others which have a tendency to spread wide and deep, which are accompanied by lesions of a far graver description, and which never fail to mar the patient's face with indelible scars. These are Lupus and its varieties, three of which are generally recognized by Dermatologists.

a. Erythematous, *Erythematous Lupus.* This variety appears on the forehead or cheek as a circular patch of shiny redness, slightly raised above the level of the surrounding healthy skin. The patch soon enlarges, spreading at its circumference while it heals at its centre. The parts attacked by it, even though there may have been no ulceration, invariably become the seat of superficial scars and assume a peculiar white smooth shiny aspect. Fine white scales cover the erythematous patch to be shed and again renewed from time to time before the skin regains its healthy condition. The disease is usually chronic and confined to the same locality, being at first no larger than a florin and gradually increasing till a large portion or even the whole of one side of the face becomes involved. Unless from some imprudence in diet or exposure to sudden changes of temperature, it is usually free from pain, heat, or irritation.

b. Non-Exedens. *Lupus Non-Exedens* appears first as a collection of little round soft violet elevations, arranged side by side so as to form patches of considerable size and usually of circular or oval shape. Though the face is its favourite locality, no part of the body can be pronounced exempt. When the elevations have attained a certain size, they are apt to remain stationary, and with the exception of slight desquamation no further change takes place.

Sometimes the affected skin and even large patches of the adjacent healthy skin undergo enormous hypertrophy and swell out to double or treble their natural dimensions. When this occurs in the face, all trace of the features becomes lost; the cheeks form a couple of huge flabby pendulous masses, and the eyes are closed up from distension of the eyelids—all which together with the enlargement of the lips and ears combines to produce frightful disfigurement. Occasionally the penis and scrotum become the seat of the disease and suffer in the same way. In a case of this kind which recently came under my notice, notwithstanding the immense size of the organ, its owner was able to continue sexual intercourse to the entire satisfaction of his partner, and, oddly enough, without discomfort to himself.

Lupus Exedens, though commencing much in the same way as the last variety, is a far graver malady in its onward progress. The tubercles deposited in the skin, instead of remaining stationary or retrograding, become inflamed and ulcerate; and, according as the ulceration spreads in width or depth, two varieties are presented to our notice.

In the first variety (which chiefly occurs in young women) fresh tubercles from time to time make their appearance by the side of the old ones and pursue the same course. A considerable area of skin is thus gradually involved, the central portion cicatrizing, while the disease is extending along the

c. Exedens.

circumference. It may last for years and in the course of that time wander over the whole face. The scar which it leaves is white, shiny and irregular, and, when occurring in the neighbourhood of the eye and mouth, causes hideous distortion of those features.

The second variety rarely attacks persons under 40 years of age, its favourite localities being the nose and cheek. If left alone, nothing stops its progress; it will gradually eat its way through skin, cellular tissue, fascia, muscle, cartilage, and even bone. Thus, when it attacks the nose, it gradually destroys all before it, till at last the vault of the palate is gone, and nose, mouth and throat form one hideous gaping chasm.

Treatment of Lupus.

Treatment.
a. Internally.

The treatment of Lupus must comprise the general dietetic, hygienic, and therapeutic measures adapted for the Scrofulous diathesis. These will be fully detailed further on (see p. 109). Cod Liver Oil in large doses and for a continued period, together with the preparations of Iodine (especially the *Syrupus Ferri Iodidi*) or of Arsenic—these are the internal

b. Locally.
remedies from which most benefit may be expected. Local treatment is here of the greatest value, for in that rodent form of the disease, the Lupus *Exedens* (or *Vorax*, as it has been aptly termed), as soon as the diagnosis is clearly established, prompt and thorough destruction of the whole ulcerated surface by means of caustics affords the only chance of arresting its progress. For this purpose we may employ the pure Nitric Acid, or some one of the so-called Caustic Pastes. There are two of them which enjoy considerable reputation as efficient and at the same time manageable caustics. (1) Vienna paste,

composed of Caustic Potash and Quicklime. (2) Mance's paste, a compound of white Arsenic, Cinnabar, and burnt sponge. A useful caustic application may also be formed of equal parts of Chloride of Zinc and starch or flour. Whichever is employed, the pain is usually very great, so tha Chloroform must be inhaled during its application, and Opium taken internally afterwards to lull the pain when the effect of the volatile anæsthetic has passed away. If the cauterization has been thorough, the eschar at the end of two or three weeks will separate and leave a healthy surface which will slowly heal by granulation. In the milder forms of the disease, such heroic treatment as this is of course unnecessary, and stimulating lotions of Nitrate of Silver, Tincture of Iodine &c. or friction with Oil of Cade, will form the appropriate remedies.

The majority of the secondary or deep cutaneous lesions in Scrofula are preceded by the primary or superficial, and in many cases the former are merely an aggravated degree or an advanced stage of the latter. Sometimes, however, they make their appearance without any previous symptoms, and, when this is the case, there is usually great intensity of the diathesis. In both cases the crusty exfoliation is an important phenomenon. Whatever may have been the primitive lesion,—whether vesicle, papule, pustule or bulla,—the result is invariably the formation on its summit of a thick brown flat or conical crust, hard to detach and composed of a number of concentric layers of epidermis. When the crust has been removed either accidentally or by force, it reveals a surface generally ulcerated and irregularly excavated, sometimes merely elevated and granular and exuding a thin purulent discharge. The appearance of a Scrofulous ulcer is characteristic. The granulations, if any, are pale flabby and moistened with a thin ichorous discharge.

<small>General characters of the deep cutaneous lesions in Scrofula.</small>

The edges are usually thin and undermined; sometimes on the other hand thickened, indurated, and everted. There is no sign of reaction. All the appearances betoken a state of constitution in which the blood is too poor, and the forces of life too low, to furnish the highly-vitalized plasma needful for the repair of structural lesion.

C. Seated in the Mucous Membranes.

Scrofulous affections of mucous membranes either Catarrhal or Eruptive.

Mucous membranes are organized in a manner precisely analogous to the external skin. There is first the superficial layer, which corresponds to the epidermis and is composed of a thin homogeneous basement membrane covered with epithelium. Beneath this comes the deep layer, which answers to the dermis and is composed of highly vascular connective and glandular tissue, varying in thickness and in secretory activity in different parts of the body. Hence Scrofulous affections of these membranes may be conveniently divided into two great classes, Catarrhal and Eruptive, according as they are seated mainly in one or other of these layers.

A. The Catarrhal.

Scrofulous Catarrh of a mucous membrane is a very simple affection *per se*, but is apt to become invested with considerable importance on account of the functional disturbance of the subjacent organ, as well as on account of the accidents and complications which may accompany its progress. The first stage is characterized by inflammatory swelling of the membrane; dryness of surface from suppression of the natural secretion; more or less itching, tingling, or burning; and lastly functional disturbance of the organ lined by the affected membrane. This stage of dryness is usually of short duration and is soon succeeded by the second stage, in which secretion returns and pours out a colourless serous fluid. The latter is of an unhealthy character and apt to irritate and excoriate the surface over which it flows. Hence it is that Scrofulous ophthalmia is so constantly attended with erythema of the eyelids and upper part of the cheeks, coryza with swelling and excoriation of the upper lip and septum nasi, and so on. In the third stage the secretion becomes opaque, puriform, and less acrid.

All this is just what occurs in common catarrh of a mucous membrane; but Scrofulous catarrh has in addition certain peculiar characters. In the first place it is more chronic in its progress and more stubborn in its behaviour with remedies. In the second place the inflamed membrane after a time loses its smoothness, becoming rough and granular. Thirdly, the irritation rarely fails to extend to the adjacent lymphatic glands, and to produce in them various degrees of inflammatory mischief. And lastly, when the inflammation has subsided, it is very apt to leave traces of its presence in the shape of slight scars upon the mucous surface, together with thickening and induration of the submucous tissue. In respect of liability to Scrofulous catarrh the following is about the order in which the several mucous membranes should be placed:— *Their peculiar characters, and effects on the mucous membranes of the*

1. The Conjunctiva, including the mucous membrane not merely of the external eye but also of the lachrymal canals. The principal secondary lesions which follow inflammation of this membrane are:—lippitudo; a granular condition of the eyelids; pustules, opacities, and ulcerations of the cornea, with more or less impairment of vision in consequence; and fistula lachrymalis. *Eye,*

2. The Pituitary membrane. Here the consequences of Scrofulous catarrh are swelling of the membrane; partial or complete occlusion of the nostrils; a nasal twang of voice; *Nose,*

impeded respiration; a swollen hypertrophied condition of the nose; and a discharge through the nostrils of purulent mucus which concretes, forms crusts, and assists the swelling of the membrane in closing up the nasal cavities.

Ear,
3. The Auricular membrane. The result here is, first of all, swelling of the membrane with chronic purulent discharge. After a time the delicate structures of the internal ear are apt to become involved, and partial or complete loss of hearing is the consequence. In young children the inflammation not unfrequently extends to the petrous portion of the temporal bone and from thence to the dura mater of the brain, giving rise to cerebral meningitis and speedy death.

Fauces,
4. The Faucial and Pharyngeal membrane. The evil consequences are—a tumid granular condition of these membranes; a gutturo-nasal voice; impairment of hearing, from obstruction of the Eustachian tube; a loud humming in the ears while talking or coughing; hypertrophy of the tonsils, which impedes respiration and gives rise to snoring and sometimes a sense of impending suffocation during sleep. The patient also suffers from a troublesome accumulation of thick phlegm about the posterior fauces; and in consequence of this the sense of smell becomes more or less impaired.

Genito-Urinary Canal,
5. The Genito-Urinary membrane. In the male, Scrofulous catarrh of the urethra gives rise to muco-purulent discharge with engorgement of the urethral glands. Prostatic enlargement and stricture are occasional, though rare, complications. When it attacks the glans, balanitis ensues; the glando-preputial membrane pours out a thick offensive secretion, becomes excoriated and, in severe or neglected cases, granular or even warty. Between Scrofulous catarrh and gonorrhœa M. Bazin maintains the following diagnostic point of difference,

viz.—that the former attacks by preference the prostatic portion of the urethra, and the latter the distal portion, especially the fossa navicularis.

In the female, there is a catarrhal condition of the utero-vaginal canal with a muco-purulent discharge, which is apt to cause excoriation of the vulva and an eczematous eruption on the inside of the thighs. When this occurs in young unmarried females, it is a complication of Scrofula much to be dreaded; for not only is it a source of much discomfort to the patient, but her relatives are too apt in their ignorance to attribute it to improper causes and, by expressing suspicion or making downright accusation, subject the innocent girl to unspeakable feelings of shame and misery. The public are of course unable to distinguish between a discharge from impure connection and a discharge from want of cleanliness or constitutional causes. Indeed they may be so much alike that it is sometimes a difficult task even for the medical man to do so, though a careful examination of the patient and consideration of all the circumstances of her case will scarcely ever leave him in doubt as to the true origin of her complaint.

6. Scrofulous catarrh of the Pulmonary and Gastro-intestinal mucous membranes I do not take into consideration, since these affections fall rather within the province of the professed Physician than the Dermatologist. *Lungs and intestines.*

With the exception of M. Rayer, Dermatologists have devoted but little attention to actual eruptions of the mucous membranes. In these membranes, as in the skin, two kinds of vesicle occur—the first identical with the conical elevated vesicle of Eczema, the second corresponding to the large flat vesicle of Herpes. Granulations on the mucous membrane answer to papules on the skin; tubercles are common to both. *B. The Eruptive.*

108   SKIN DISEASES

Pustules are by no means rare on those mucous membranes which are in close contiguity to the skin, but they differ from their kindred in the latter situation by their tendency to early rupture. In a word, nearly all the Scrofulous eruptions, which occur on the external, may and do occur on the internal skin or mucous membrane.

Treatment.
a. Internally.

The most common exciting cause of these affections being exposure to cold and damp, the first desideratum for the patient will be avoidance of these pernicious influences and, if possible, residence in a warm dry bracing locality. Another common cause, in persons predisposed to them, is over-use of the organ which the mucous membrane subserves, as for instance when the conjunctiva becomes inflamed by straining the sight too much with continuous reading, especially when the reading is done by the glare of artificial light. Besides

b. Locally. the general treatment proper for the Scrofulous diathesis, which will presently be detailed, certain local measures may be adopted with great advantage. The utmost cleanliness must be enforced to remove from time to time the unhealthy muco-purulent discharge and prevent its injurious effects both on the membrane itself and on the adjacent sound skin. It is not sufficient merely to wipe or wash the membrane; many parts of it (as for instance the sinuses and windings of the membranes of the eye, nose and ear) will be inaccessible to ordinary manipulation of this kind. The only way thoroughly to cleanse the whole diseased surface is to syringe it frequently with a steady and well-directed stream of tepid water or milk and water. Lotions, astringent, stimulant, caustic, detergent or deodorizing, may be injected, according as the condition of the membrane and the character of its secretion may indicate. In obstinate cases, counter-irritation to the skin in the neighbourhood of the inflamed membrane should never be neglected. The application from time to time of a small blister to the temple in Scrofulous ophthalmia, or behind the ear in Scrofulous otorrhœa, is generally attended with the best results. In the treatment of discharges in the female, if there is good reason to suspect the existence of ulceration in some portion of the utero-vaginal canal, the speculum is necessary not merely to clear up the diagnosis but also to facilitate treatment. In Scrofulous females the os and cervix uteri are sometimes the seat of ulceration of a most destructive kind. That formidable disease, the so-called 'rodent ulcer of the os' is to the womb exactly what Lupus Exedens is to the face.

GENERAL TREATMENT OF THE SCROFULOUS DIATHESIS.

The general treatment of the Scrofulous diathesis may be considered under the three heads of hygiene, diet, and medicine.

Hygiene.

For all human beings, but more particularly the subjects of Scrofula, the first and foremost condition of health is abundance of light and air. It may seem almost superfluous to state formally such a truism as this, but experience abundantly shows that it is one thing to believe an abstract truth, another, quite another, thing to carry it into practice. How common is it for families, who possess large airy front rooms in their houses, to shut them up except for receiving company, and dwell habitually in gloomy cramped back rooms! Again, how commonly do people, who insist on having spacious airy rooms to live in by day, consider any little stuffy chamber good enough to sleep in by night! Yet fresh air and cubic space are not less indispensable during one period of the twenty-four hours than during another. Among atmospheric influences,

Light and Air;

next to want of light and air, nothing is so powerful in evoking the Scrofulous diathesis, and aggravating it where it already exists, as damp and cold; not cold alone, but cold and damp combined. Hence patients with Scrofulous taint or actual Scrofulous disease should, if possible, take up their residence in a mild dry but at the same time bracing locality. They need not for this purpose expatriate themselves to the South of France, the Coast of the Mediterranean, or to Madeira; plenty of places are to be found in our own favoured island. Observation has long ago disproved the necessity or even the desirability of a warm climate for a Scrofulous patient. What he needs is not so much warmth as evenness of temperature, in other words a temperature free from sudden and great vicissitudes.

Kind of climate desirable;

Daily ablution of the whole body and subsequent friction of the skin with a rough towel are a grand means of preserving health in this disease. Perhaps of all the so-called anti-Scrofulous hygienic remedies sea-air and sea-bathing are the most powerful. They impart a tone to the system which cannot be attained by any other known means. With Scrofulous patients, however, bathing requires to be practised with certain precautions. As a rule, they should not bathe in the early morning or on an empty stomach. The forenoon, when the system has been fortified with breakfast and the sun has dispelled the mists and clouds, is the most favourable time. A cold water plunge before breakfast with the walk to and fro is rather a tax upon the bodily strength; and a Scrofulous patient, if he attempts it, is generally more or less exhausted and below par for the remainder of the day.

Ablution and friction of skin;

Sea-bathing;

Cheerful games and occupations entailing active exercise in the open air, both in summer and winter, are most desirable. With Scrofulous patients there is a special reason for this.

Open-air exercise.

Such exercise stimulates their sluggish circulation and counteracts that tendency to coldness of the extremities to which they are so liable, and which, by causing the blood to accumulate in the internal organs, is so common a cause of visceral congestion and inflammation. Habits of prolonged study, sedentary trades and occupations, cannot be too strongly deprecated under any circumstances in this diathesis, but more particularly when the body has not yet completed its growth.

The clothing of the patient is a point on which a good deal might be said. Scrofulous subjects, especially children, should be clad more carefully than they usually are. They are often allowed to dawdle about in cold weather with necks and arms bare, and their legs exposed from the ankle up to or even above the knee. This is a great mistake. The young (and old) have a comparatively limited power of resisting cold, in other words, of maintaining an independent temperature. Parents should therefore be very cautious about exposing their children to cold just at that time of life when such exposure can be least tolerated. By all means keep children in the open air as much as possible, even in the coldest weather; but not without adequately protecting every portion of their trunk and limbs. Flannel is undoubtedly the best material to wear next the skin. It should be worn both summer and winter, and in delicate chilly subjects by night as well as by day, for during sleep the vital powers are weak and the susceptibility to cold is proportionately strong. *Clothing.*

Diet.

As to the proper diet for Scrofulous subjects, no precise rules can be laid down. The diet must vary with the circumstances of the case. Not unfrequently we have to perform the *Food.*

very ticklish task of adapting it to the caprice of our patient's appetite as well as to the requirements of his disease. Perhaps the error most needful to guard against is that of restricting the patient, or allowing him to restrict himself, to too monotonous or too thin washy a diet. In spite of much that has been said and written to the contrary by *doctrinaires* of learning and distinction, I firmly believe the truth to be this; that a patient may be safely left to follow pretty much the dictates of his appetite, provided of course that he eat in moderation and that the diet of his choice contain the necessary elements of nutrition.

Alcohol. With regard to the use of alcohol in this disease, I agree neither with those who altogether withhold it nor yet with those who would keep the patient in a state of chronic 'fuddle,' with a daily skin-full of wine, beer, or porter. If by taking a moderate quantity of wine or beer a patient finds his appetite and digestion improve, by all means let him continue to take them. If on the other hand they upset digestion and so cause less food to be assimilated, the sooner their use is relinquished, the better.

Medicine.

Whatever may be the local lesions in this disease, they must never be allowed to divert our attention from the state of the general health. Mal-assimilation of food and consequent faulty nutrition are the first links in that series of morbid processes, local and general, which constitute the Scrofulous diathesis. They are, so to speak, poisons which pollute the stream at the fountain-head; and it is against them from first to last that our efforts must be unceasingly directed. There is no known specific against Scrofula. The grand specific, the omnipotent Panacea, to be striven after in this and in all other diseases of the kind, is a healthy condition of the patient's own blood.

Mal-assimilation of food the great evil to be remedied.

No known specific against Scrofula.

But to manufacture healthy blood, we must have a healthy appetite and a good digestion—the two blessings of all others which least often fall to the lot of a Scrofulous patient. Capriciousness of appetite, loathing of animal food especially of fat, together with the various forms of dyspepsia, are with him the rule, not the exception. Our chief medicines in combating these symptoms will be the different vegetable bitters, the alkalies, mineral acids, and gastric sedatives, separately or conjointly as the circumstances of the case may require. As a general rule it will be found that, where there is a foul tongue and irritable stomach, alkaline preparations (such as the Carbonate of Magnesia, Bicarbonate of Potash &c.) do most good; while simple loss of appetite without much gastric derangement is benefited most by the mineral acids. Purgatives cannot be dispensed with, but they should be of the mildest, least irritating description. Cases of habitual constipation often admit of relief by adopting a somewhat coarser or more varied diet. If the addition of fruit, fresh vegetables &c. to the ordinary diet and the substitution of brown bread for white do not suffice, it will be far better for the patient to use a tepid or cold water injection than to be physicked two or three times a week with aperient pills or draughts. *Therapeutic value of vegetable bitters, alkalies, acids, &c.* *Of purgatives,*

Although, as I said above, there is no known specific for Scrofula, there is one medicine which, if the patient can take it, generally does more good than any other, and that is Cod Liver Oil. The rapidity with which under the use of this oil morbid symptoms vanish and patients regain appetite, colour, flesh, and strength, is often marvellous. The oil should be taken on a full stomach, the dose being at first small and gradually increased. The flavour, if disliked, may be effectually disguised by the addition of a minute quantity (about *Of Cod Liver Oil,*

I

$\mathfrak{m}\tfrac{1}{4}$) of the Essential Oil of Bitter Almonds. The appearance as well as the flavour may be masked by mixing it with equal parts of Lime Water; a milky emulsion is thus formed and is specially adapted for cases where acidity of the stomach coexists. By having recourse to such contrivances as these, few cases will occur in which the patient will be unable to tolerate the Cod Oil. If however he cannot take it even in very small quantities, cream will often be found to agree and to answer the same purpose. The creams of Somerset and Devon are particularly suited for the stomach of a Scrofulous patient on account of their extreme delicacy, lightness, and ease of assimilation.

<small>Of Cream,</small>

The preparations of Iodine, Iron, and Cinchona Bark, rank next in importance to Cod Liver Oil. The Iodide of Iron in the form of Syrup is an invaluable medicine in the treatment of most Scrofulous affections, whether external or internal. Iron to do good should be given in a readily soluble form and in small doses, otherwise it is apt to upset the stomach and cause some little febrile disturbance. With Scrofulous patients Quinine agrees better when taken in pill than in mixture, and on a full than on an empty stomach.

<small>Of Iodine, Steel, Bark, &c.</small>

The remedies which have been and still are vaunted as possessing wondrous efficacy in the treatment of Scrofulous affections are innumerable, but those above mentioned are almost the only ones which have stood the test of experience and therefore may be deemed worthy of mention.

www.ingramcontent.com/pod-product-compliance
Lightning Source LLC
Chambersburg PA
CBHW031345160426
43196CB00007B/734